Papermaking in Eighteenth-Century France

Thiers
To: Paris
BURGUNDY
LYONNAIS
Lyons
Rhône
Ambert
AUVERGNE
Voiron
Rives
Deûme
Vidalon-le-Haut
Annonay
Cance
Grenoble
Tence
DAUPHINÉ
Privas
Rhône
LANGUEDOC
N
W
E
S
PROVENCE
Uzès
KILOMETERS
0
10
20
30
40
50
Rhône
Rhône Valley
circa 1789
Montpellier

Papermaking in Eighteenth-Century France

Management, Labor, and Revolution at the Montgolfier Mill

1761–1805

Leonard N. Rosenband

The Johns Hopkins University Press
Baltimore and London

Published 2000
Printed in the
United States of America
on acid-free paper
2 4 6 8 9 7 5 3 1

The Johns Hopkins University Press
2715 North Charles Street
Baltimore, Maryland 21218-4363
www.press.jhu.edu

Library of Congress
Cataloging-in-Publication Data
will be found at the end
of this book.

A catalog record for this book
is available from the
British Library.

ISBN 0-8018-6392-9

TO REVA

who gave me

my voice

Contents

Preface

Fine paper is still made in a mill just above the French town of Annonay. Today, these sheets are fashioned by machines governed by computers. Nearby these modern devices lies a dank basement chamber where women once sorted rags, the raw material of handmade paper. I first saw this hollow, which now contains little more than debris, on a memorable tour of the Canson and Montgolfier paper mill conducted by M. Xavier Frachon, its former manager. Throughout the mill other traces of papermaking's premechanized legacy abound, most notably the precious, carefully wired molds on which artisans produced reams of paper with a customary set of motions. This book is about that distant world of work and even more about the regulation of that labor.

Adélaïde de Montgolfier never forgot the sounds of Vidalon-le-Haut, her family's principal paper mill and one of the largest in eighteenth-century France. Its product had established the Montgolfier name long before balloon flights earned international fame for her father and uncle. Many years later, she remembered the roar of Vidalon's stamping mill, the shriek of the foreman's whistle, the songs sung by the rag-sorters to their children, and the whisper of the wind in nearby willows and poplars.[1] The machine and the garden were in harmony.

But Mme Montgolfier failed to hear Vidalon's discordant notes. Eight years before the French Revolution, the mill was the setting for a bitter strike and successful lockout. This struggle, one of many labor disputes that bedeviled French papermaking during the twilight of the Old Regime, was sparked by the collision of two conceptions of the proper order (*bon ordre*) of the craft. Both master papermakers and skilled journeymen sought to regulate their trade, to shape the social relations of production in accord with their needs and values. The result was industrywide turmoil and, at Vidalon-le-Haut, a new system of work discipline.[2]

This book explores the sources, design, and outcome of an early experiment in industrial labor discipline. It is a study of the Montgolfiers' pioneering effort to transform the balance of power between masters and men in their own favor. Labor discipline, as Vidalon's *patrons* understood it, consisted of

creating fresh terms and a novel environment for the workers' exercise of their skills.[3] To do so, the Montgolfiers attempted to exile the journeymen's custom from their mill and to snap the transmission of skill from father to son. To replace their peripatetic veteran hands, they trained a flock of young men and women—a generation, they dreamed, of malleable employees.

Blueprints and schemes, however, are not the measure of shopfloor realities. Indeed, the provisions of many disciplinary codes can be read as catalogs of common evasions of the bosses' wishes. Further problems of conception and evidence have haunted the consideration of past programs of labor discipline. There has been a tendency to embrace simple models of adaptation and resistance, as if clear thresholds of worker "commitment" existed—or would be noticed by the employers who always wanted more.[4] That the man who earned his master's approval during the day might draft an anonymous, threatening letter or burn a hayrick at night reminds us of the many dimensions of individual conduct, as well as the opacity of such men to their betters. Certain codes and regimes, which appeared airtight on paper, dissolved as the strong hand—and cane—of an energetic figure (like Josiah Wedgwood) disappeared.[5] Finally, the place of the deeds and desires of workers themselves in shaping the disciplinary process needs additional emphasis. After all, much of the Montgolfiers' design was a direct response to challenges posed by their journeymen.

Perhaps the best strategy to circumvent these obstacles in the study of work discipline is to shrink the compass to a particular industry and specific workshops within it. The distinctive features of papermaking framed the Montgolfiers' program of labor discipline, as well as the conflict that animated it.[6] Above all, the production relations of the industry had always been capitalist. A millmaster, generally acting in the name of a stationer or absentee owner, made the essential daily decisions about the use of equipment, raw materials, and labor. A mobile, skilled corps of journeymen survived on their wages and wits. They sweated in mills that varied from petty shops counting no more than eight or ten hands to grand enterprises of more than one hundred, such as Vidalon-le-Haut. But the division of labor and the general pattern of production differed little from small mills to large. Theirs was a craft, the Montgolfiers knew, hidebound with centuries of familiar technique and customary practice. Some of these conventions ensured that the reams would be composed of fine, white sheets; others disrupted production, cemented worker solidarity, and enabled the journeymen to police the labor market. To the Montgolfiers, the workers' ingenuity had forged the shackles of "blind routine." As they remodeled Vidalon-le-Haut into living *Encyclopédie* plates, they envisioned workshops enlightened by efficiency and the spirit of in-

novation.[7] Still, the Montgolfiers did not have an alternate division of labor at their disposal, nor did they imagine that the sweeping mechanization of their trade was imminent. To establish the proper setting for the "perfection" of their wares, they intended to uncouple the journeymen's skills from their cultural moorings. The events of 1781 provided them with the opportunity to do so; this book assesses the results of that attempt.

Finally, this study investigates a singularly important transition that rarely receives the attention given to the journey from field or workshop to factory: the transformation of already elaborate production relations in a mature industry undergoing limited technological change. It bridges the historiographical divide that separates the social history of work and the social history of the management of that work. Only by linking the two approaches can we map the intricate terrain of industrial organization between handicraft and mechanized factory. Only then can we interpret what the Montgolfiers—stern Pierre, calculating Etienne, rambunctious Jean-Pierre—and their workers—loyal Payan, mutinous Deschaux, besotted Joseph Etienne—made of their trade, their labor, and each other. And only then can we hear the discordant notes that eluded Adélaïde de Montgolfier.

Acknowledgments

I have acquired many debts during the years that I worked on this book. M. Xavier Frachon was a gracious host and delightful guide while I conducted my research. The staff of the Archives nationales, the Archives départementales de l'Ardèche (Privas), the Archives départementales de l'Hérault (Montpellier), and the Bibliothèque municipale de la ville d'Annonay were helpful beyond the call of duty during my frequent visits to their collections.

Fellowships have aided me at critical moments. I am grateful to the American Council of Learned Societies (1982), the Shelby Cullom Davis Center for Historical Studies at Princeton University (1995), and the Dibner Institute for the History of Science and Technology at the Massachusetts Institute of Technology (1996) for unfettered hours of writing and reflection.

Robert J. Brugger and Norman Prentiss at the Johns Hopkins University Press have made the final stages of publication a pleasure. I am grateful for their patience as well as their astuteness. Wendy Jacobs was an attentive, sympathetic copyeditor. The engravings in chapter 2 were provided by the Robert C. Williams American Museum of Papermaking. I am indebted to this institution for permission to reproduce them and especially to Cynthia Bowden for her help on short notice.

Many friends, colleagues, and teachers have offered valuable insights, shared information, and provided encouragement along the way: Timothy Barrett, Perry Blatz, Roger Chartier, Robert Chesler, Gary Cohen, Natalie Zemon Davis, James Farr, Laura Fisher, William French, Anthony Grafton, William Jordan, John Krill, Eric Olsen, Carlo Poni, William Sewell Jr., Daniel Vickers, Anthony F. C. Wallace, and Benjamin Weiss. Christopher Johnson has been a tough-minded mentor and an engaging companion; Daryl Hafter has asked wonderful questions in her gentle way; and Jeff Horn has blended criticism, candor, and congeniality. Pierre Claude Reynard, a fellow student of Old Regime papermaking, has been an open, invaluable colleague, swapping manuscripts and archival tidbits with me.

At Utah State University, my employer for fifteen years, I wish to acknowledge the support of Robert Cole and Norman Jones. Michael Nicholls and I

have shared many conversations about the eighteenth century, and he has been kind with my early formulations. David Weiland, a colleague of recent vintage, has sharpened my thinking about economic matters. Barre Toelken never let me forget about culture in the lives of paperworkers. And Ross Peterson stood by this project—and me—loyally over the years.

Robert Darnton supervised the doctoral dissertation on which this study is based. He did much to help me clarify my ideas and, even more, to express them. Charles Gillispie, trusted adviser and generous friend, has shared with me his unsurpassed knowledge of Old Regime science and technology. Our letters, conversations, and hours devoted to the Montgolfiers inform every page of this book.

Brad Gregory, once my student and now my colleague and friend, listened patiently to my ideas and sentences, and made them better. Of all the things Utah State University has provided to me, none has been more important than my friendship with Brad.

Thomas Safley, cherished friend and intellectual companion par excellence, knows how much he contributed to the maturation of this manuscript and to the refinement of my thinking about the European past. Now I can finally offer public thanks for so much private encouragement.

Carolyn Doyle typed the manuscript with skill and gentle good humor. I doubt that she will ever realize all that she did to speed the appearance of this book.

Ira and Shirley Jaffe, my in-laws, have offered quiet yet valuable support since my days as a graduate student. Maurice Rosenband, inveterate storyteller, taught me to love a good narrative; Regina Rosenband, with her boundless compassion, taught me about equity and justice. Together, my parents never let my commitment to the study of the past flag.

My daughter, Amy, has heard about the Montgolfiers throughout her life. Her sweet cheer has buoyed me when I needed it most. Reva has provided love and instilled confidence in equal measure—no man could ask for more.

Money, Weights, and Measures

One *livre tournois* = 20 *sous*
One *sou* = 12 *deniers*

The Montgolfiers marketed their paper in Lyonnais *livres* (pounds) of 14 ounces.

A ream of paper included 500 sheets; a post of paper, the production measure, varied according to the weights and dimensions of the paper.

PART ONE

An Old Industry

CHAPTER ONE

French Industry in the Eighteenth Century

The smooth grain and robin's-egg-blue tint of Dutch paper posed enormous problems for French manufacturers. It cost them markets, shuttered their mills, and revealed their technological inertia. To compete with the Dutch, the papermakers claimed, their industry had to be deregulated. But it was not the technological tutelage or strict prescriptions of a Colbertian state that had to be cast off.[1] Rather, it was the journeymen's capacity to conserve immemorial techniques and withdraw their labor at critical moments that provoked the masters. They petitioned the state for subventions to emulate Dutch practices and for remedies to domesticate their ungovernable hands.

During the last decades of the Old Regime, Peter Jones has argued, the French state fairly hummed with reform.[2] It employed technological travelers who ferreted out the secrets of foreign rivals, including Dutch papermakers, and helped establish them at home. A second strand of reform would subordinate workers and journeymen in every trade, especially by tying them down. This was the age of the certificate of good conduct and the first *livret*, the documents that would enable employers to fish only compliant, assiduous men from the streams of itinerants. Still, there was a sense, as the hatters put it in 1764, of a particular boldness, of a growing "taste for independence" among journeymen throughout the crafts.[3] Reams of royal edicts and regulations had yielded neither stable nor tractable paperworkers. Nevertheless, the Montgolfiers banked on state aid to retool their mill along Dutch lines. In the course of refashioning their work force, they would turn to the counsel of a state official and, even more, to the millwright-draftsman he dispatched to Annonay.

The reformist state, the convictions of journeymen sure of their skills and ways, and the uncertainties of an art dependent on limpid streams and short supplies of old linen, all influenced the nature and force of the Montgolfiers' approach to work discipline. So, too, did their standing in the craft, their confidence in applied science, and much more. Accordingly, Vidalon's *nouvel ordre* was neither a simple vector of technological advance nor the embodiment of a stage in the evolution of the organization of production. It was an

attempt to create an unfettered domain for innovation. And it was an effort to train a cadre of workers more responsive to the Montgolfiers' sticks and carrots than those wielded by their skilled brothers.

If the pace of French industrial activity quickened on the eve of the French Revolution, it did so mainly in traditional arteries. While England was embarking on its career as the workshop of the world, France did not experience an industrial revolution. For every John Holker, large-scale cotton manufacturer and royal bureaucrat, or Jacques de Vaucanson, inventor of an automatic silk loom, countless men and women employed time-honored production processes in cluttered ateliers. In 1789, Great Britain possessed more than twenty thousand spinning jennies and two hundred mills of the Arkwright model; France had less than one thousand jennies and eight of these mills.[4] The diffusion of new techniques in France was impeded by aggressive loyalty to the familiar: even the few French factories, Pierre Léon observed, "conserve[d] a rustic appearance, which was similar to an artisanal shop."[5] Maintaining the equilibrium within each community of craftsmen was an ideal strong enough to survive the demise of the guilds in 1791. Antoine-François Montgolfier fretted about the harmful potential of the technological advantages enjoyed by his nephew, Etienne, because "he can lower the price of his papers, and wipe out the commerce of his *confrères*."[6] The French entrepreneur who constructed his fortune over the remains of others, David Landes wrote, was a *"mangeur d'hommes"*—a cannibal.[7]

French industry, then, was characterized by an early start, gradual alterations in technology and business practices, limited mechanization and the continued preeminence of handicraft production, and an emphasis on quality wares and individual design rather than cheap, standardized goods.[8] Still, changes were in the works. Léon maintained that the general enrichment and demographic expansion of eighteenth-century France had fattened its internal markets.[9] (Colin Jones has even detected a mini-consumer revolution, which Cissie Fairchilds has attributed to "populuxe" goods, less expensive knock-offs of luxury wares intended for the middling and popular classes.)[10] Overseas commerce, too, had stimulated manufacture in towns, cities, and their hinterlands. Liberal ideology increasingly challenged the rights and privileges of the craft communities, while the extension of industrial activity into the countryside undermined the effectiveness of guild wardens.[11] Certain entrepreneurs displayed a newfound willingness to emulate advanced techniques, even if they were British, Swiss, or Dutch.[12] A pattern of unbalanced growth emerged during the century as relatively new industries, such

as calico printing, took off, and venerable mechanical arts, including glassmaking and papermaking, shook off organizational and technological cobwebs. Even guildsmen engaged in subcontracting, employed "illegal" hands in hidden garrets, and formed partnerships with men linked to neighboring craft communities or their own journeymen.[13] It was in the midst of this "age of manufactures," when craft customs and skills mingled with new merchandise, new attention to fashion, and new approaches to production, that the Montgolfiers devised their nouvel ordre.

Historians now recognize that the transition to modern industrial capitalism was "a process that fostered a *variety* of possible forms of industrial organization."[14] Seemingly contradictory phrases, such as "factory artisan," reflect our sharpened yet unsettled perception of early industrial circumstances. Papermaking shared this category-blurring quality. During the eighteenth century, it still consisted of a series of centuries-old artisanal procedures carried out in a mill or a consolidated group of workshops under centralized control. Capital-intensive (by the standards of premechanized production) and geared to the fabrication of lots, or batches, papermaking always entailed elements of supervision.[15] After all, whether the reams were fine writing paper for courtiers or gray sheets for local consumption, they had to meet the requirements of the market. Small wonder that millmasters sought steady attendance, application, and output. The journeymen, of course, recognized these imperatives, and negotiated accordingly. They did so on the basis of their skills, which they couched in custom, the better to keep them scarce.

This book, then, is history from the workshop up. It explores how paper was made, the tools, techniques, raw materials, and seasonal flows of a venerable trade. It examines how this work was accomplished, the conflicts and compromises between masters and men. And it probes the strategies devised by both the Montgolfiers and their workers to deal with it all. In doing so, it reveals discipline as it was lived: an unexpected bonus, a harsh rebuke, the flight of a favored hand, or the production of shoddy wares in pursuit of an incentive.

Still, the Montgolfiers and their journeymen also inhabited a broader craft, itself embedded in a particular political economy. Vidalon's nouvel ordre took shape in the last years of the Old Regime, when the state—at least as regards papermaking—sustained the Colbertian legacy of technical virtuosity while largely discarding the same tradition's regulatory remnants. This balance was now expressed in an Enlightened idiom; its watchwords were instruction rather than imposition, prompting and premiums rather than prescription. Moreover, a growing recourse to applied science and its adepts by

entrepreneurs and government alike linked the two in bonds of policy and action. Such were the ties that emboldened the Montgolfiers, just as a state subsidy aided the remodeling of their mill.[16]

As the demand for fine paper rose, the supply of old linen remained scarce, and the challenge of refined Dutch practice, with its handsome, competitively priced product, persisted. These pressures figured in the Montgolfiers' decision to mimic Dutch devices and techniques, as did the bounty from the state and the sureness inspired by applied science. Still, Vidalon's *patrons* chanced more, for they longed to propagate skill on their terms alone. Thus their nouvel ordre. Like the pioneering disciplinary regimes sketched in a classic essay by Michelle Perrot, it involved close surveillance, integrated hands through paternalist practices, and incorporated worker families in the effort to create a devoted labor force.[17] But similarity of tactics should not mask the originality of the Montgolfiers' strategy. They intended to transform Vidalon-le-Haut into a theater of experiment, a testing ground for both men and machines. With that accomplished, they would be able to match Dutch paper and secure their profits.

Such experiments remind us that transitions within the face-to-face world of early industrial capitalism were lived, negotiated, and at Vidalon-le-Haut, consciously designed. These shifts need not have embraced the entire trade, although the Montgolfiers believed they were paving the way for their fellow papermakers. Above all, the Montgolfiers' version of a technologically enhanced, employee-staffed, properly ordered mill—their capitalist utopia—did not subject workers to the rhythm of the machine.[18] The Hollander beaters, the heart of the technology they imported, transformed the pulp, only touching lightly the work of producing the sheets. Yet, dependence on their workers' skills did not prevent Vidalon's *patrons* from sharing Wedgwood's fantasy of a labor force that responded as a single set of hands. They would do so, however, as appendages of a design, rather than as appendages of the machine. The rub, of course, was that this scheme rested not only on "scientifically" determined tasks, precise work schedules, and the confines of a mill village but also on the workers' skills, their dexterity and know-how, their very humanity.

Here, then, was an elaborate plan to undermine the paperworkers' alleged (by both their masters and the state) government of the craft and, more particularly, to wrest control of their skill from them. Yet, while the Montgolfiers may have dreamed of workers as "abstract labor power" to be molded as they saw fit, they dutifully recorded the habits and attitudes of each of their post-lockout hands. That was a realistic practice, since their nouvel ordre was shaped, in part, by daily give-and-take with their old workers and new. (Not

surprisingly, Vidalon's *patrons* eagerly awaited real automatons, in the form of a papermaking machine.) Meanwhile, the Montgolfiers struggled to instill a fresh sense of direction in their young hands. They would remain at the mill, sweat over their assigned tasks, and stay apart from the tramping journeymen and their ways. How they and their bosses actually carried out this project is the story of this book.

CHAPTER TWO

Making Paper

For Michael Sonenscher, custom "supplied a vocabulary of particularity" for the craftsmen of pre-Revolutionary Paris. Through it, they excluded interlopers, petitioned for higher wages, and resisted productivist pressures. Moreover, Sonenscher contends, custom permitted men who worked with the same raw materials and shared broadly similar skills to craft separate identities and guilds.[1] Certainly, journeymen paperworkers found a vocabulary of distinction in their custom. But they turned to it as the expression of a unique set of skills—skills that had endured since making paper by hand took its modern form in late medieval Italy.

In the eighteenth century, an informed observer concluded that paper was "an everyday merchandise."[2] The perfection of movable type and the passions of the Reformation had stimulated papermaking in many parts of Europe. Yet, a painter or engraver from Luther's day would have found much that was familiar in the paper mills of Rousseau's. He would have seen women sorting rags, water-powered mallets macerating old linen, vatmen "promenading" newly minted sheets, couchers transferring the fresh paper onto dampened felts, men tugging at the bar of the great press, the layman separating still-moist sheets from the felts, women hanging the paper to dry, a sizerman stirring his cauldron of finish, more women inspecting the freshly sized paper for flaws, a teller counting quires, the loftsman wrapping reams, and swarms of children performing odd jobs. It was as if an assortment of guilds—for the tasks of papermaking were that distinct and specialized—had been yoked together under one roof. This integration placed a premium on the workers' attention to time and proper handling of raw materials. Above all, the fragile, easily soiled product itself amplified the need for precision and heightened the uneasiness of millmasters who depended on the skills and sobriety of the tramping journeymen who knocked on their workshop doors.

The sound of creaking carts, burdened with crates of old rags, signaled the beginning of work. Papermakers reputedly prized Burgundian rags above all. Even this linen, however, which was usually washed before shipment, required

careful scrutiny. Crates of spoiled or polluted rags were commonplace: in 1795, for instance, the Montgolfiers complained bitterly about one cartload of discarded linen that was "overloaded with strawdust and little pieces of wood," streaked with vomit, and marred by *sabot*-prints.[3]

At Vidalon-le-Haut, porters hauled the clammy rags to an equally humid subterranean chamber. There, rag-sorters (*trieuses*) cut knots and irreparably stained fibers from the linen and then placed the valuable remnants, according to their whiteness and other qualities, in one of six sections of a large chest. If the work of the trieuses failed to meet the Montgolfiers' expectations, the women would "receive nothing and they will be required to sort [the discarded linen] again."[4] In his *Art de faire le papier*, Joseph-Jérôme Lefrançois de Lalande reasoned that the "care, discernment and precision" demanded of rag-sorters led millmasters to rely on "responsible adults."[5] But the Montgolfiers, anxious to put the children lodged in mill housing to work, and Vidalon's female hands, eager to earn a few extra sous, pressed children into service in the sorting chamber. In 1761, Nanon Joubert, a second woman named Vinette, and their young daughters toiled together among heaps of Burgundian linen.[6]

After the rags had been washed and sorted, they were left to soak for about ten hours in a closed, cylindrical hollow known as a *pourrissoir*. The millmaster then had the fermenting linen stacked in piles, where it usually remained for a few weeks. This phase ended when a worker could no longer leave his hand inside the hot mound of rags for more than a few seconds. Such practiced judgments accompanied every step of manual papermaking and determined the quality and value of the reams: a premature end to fermentation produced a coarse, light, or stiff paper; and too many hours in the stinking heaps diminished the yield of the old linen and weakened the paper.[7]

The rags were then cut into inch-long filaments, washed, and brought to the stamping hammers. As the millwheel turned, it drove an axle in a rotating motion, which lifted and dropped a series of shafts. Each shaft ended in a mallet beveled to fit neatly into a trough. In most mills, a first group of hammers performed the gross pulping, a second tandem acted as beaters, and a third brushed or refined the filaments.

Experienced hands, too old and shaky to fabricate paper, generally supervised the washing, sorting, rotting, and pulping of the rags. The Montgolfiers expected these veterans, known as governors (*gouverneurs*), to have two days' worth of cut rags on hand at all times, to keep the hammers and troughs clean, unclogged and well swept, and to ensure the regular beat of the mallets.[8] The governors had to pay attention to the smallest change in the roar of the mill

room. Lalande marveled that these old hands slept through the routine crash of the mallets yet rose at the sounds of storm-quickened blows in time to prevent the flow of muddied water from soiling the pulp.[9]

The governor resorted to time-honored practices when he had to decide if the pulp was ready. He grabbed a handful of stock, squeezed out the water, and broke the mass open. If the filaments were homogeneous and resembled the short, flattened, and hairy appearance of a fly's leg, the governor passed them to the vatman (*ouvrier*).[10] As he dumped the pulp into the warm water of his vat, the vatman made judgments, too. More fiber and less water in his vessel, which measured 5 feet in diameter and 30 inches in depth, yielded heavier reams. The color of the water was the vatman's yardstick, and a great deal turned on his decision to thicken or thin his brew. At 35 pounds per ream, the paper known as *bâtard* became the elegant base of musical compositions; at 28–30 pounds, it was less valuable writing paper.[11]

The vatman manufactured a sheet of paper by dipping his mold, a rectangular wire mesh bounded by a wooden frame, into the vat. This unwieldy apparatus was strung with two sets of wires, thick "chain" lines that ran from the top of the mold to its bottom and supported the fine, perpendicular "laid" lines. Handmade paper bore faint impressions of these laid lines, so the Montgolfiers expected their vatmen to inspect the stringing of their molds no less that twenty times each day.[12]

Not surprisingly, the men who made the molds were among the most highly prized—and highly paid—workers in the industry. Known as *formaires*, they wandered from mill to mill, repairing old molds and wiring new ones. The Montgolfiers hired a number of these footloose carpenters during the 1780s and had one, d'Henri, in residence for more than twenty years. A reliable formaire was a valuable man; after all, the molds cost the Montgolfiers from 22 livres 12 sous to as much as 51 livres 4 sous apiece, the equivalent of about 7 to 15 percent of a vatman's annual wage at most mills.[13]

The French state also turned a careful eye to the work of the formaires. It specified precise standards for the weights and dimensions of paper produced in the kingdom and required the formaires to weave several indicators into every mold. In theory, the name of each manufacturer, the location of his mill, the measurements and quality of the paper, and the date of production should have been easily visible.[14] In practice, French *fabricants* banished watermarks from their wares, borrowed their competitors' marks, or toiled with their neighbors' molds, illicitly obscuring the source of their reams.[15] For their part, the Montgolfiers commonly sent lightweight, illegal reams to reliable customers on approval.[16]

Once the vatman had lifted the mold, with an infant sheet clinging to the

wire mesh, he began a series of traditional actions. To ensure an even product, he "promenaded" the mold, that is, he shook it from right to left and then from left to right. To prompt the fibers to knit (or "shut," in the journeymen's lexicon), the vatman then shook the mold to and fro. The right-hand edge of his mold was known as the hand; the left side of the apparatus was called the foot. *La mauvaise rue* lay against the vatman's gut, while the opposite margin, where the paper was thought to be more firm, earned the fanciful name of *bonne rive*. Most notably, the journeymen claimed that if the vatman had failed to reinforce the *bon coin*, the good corner that linked the bonne rive and the right edge, the sheet would fail to stand up when the layman "pinched" it in order to separate it from the felt.[17] Thus, the journeymen's picturesque terminology amounted to a glossary of technique. It reflected their sense of sharing the intimacies of skilled work as well as the largely static nature of their art.

The vatmen toiled according to customary workloads, which varied with the weights and dimensions of the paper. In Angoumois, they were expected to fashion 3,900 sheets of ordinary printing paper per day.[18] Despite this burdensome standard, the emphasis was not always on speed. On the contrary, many manufacturers feared hasty work by journeymen eager to make the most of piece rates. Even the slightest slip ruined a sheet—no minor matter in a craft where a debilitated vatman was said to have lost his "shake." Like the familiar elements of the mold, common flaws had their sobriquets: wrinkles were "goat's feet," splashes caused "chestnuts," and uneven swells of pulp were *andouilles*, "sausages," or perhaps "shits."

After four or five seconds on the mold, the filaments began to bind. The vatmen then removed the deckle, the wooden lid he had placed around the mold to prevent the pulp from spilling over the edge. Next he passed the mold to the coucher, whose main tools were a small easel and a pack of hairy felts. The coucher needed sure hands, since he had to "couch," or flip, seven or eight sheets per minute from the molds to the felts. If the easel, on which the felts rested, moved or his hands were unsteady, the coucher might trap air between the new sheet and the fiber, producing blisters.[19]

The coucher and the vatman worked in tandem. As the vatman slid one mold to his comrade, its twin, shorn of its fresh sheet, was on its way back to him. Their labor, rich in routine and considered choices, was exhausting. At some mills, vatmen and couchers commonly switched jobs during the day to ease the strain on backs and shoulders. Red arms, missing fingernails, respiratory ailments, and rheumatism accompanied the veteran hand's pay; indeed, journeymen paperworkers over the age of forty had to prove that they were not superannuated.[20]

The pile of fresh sheets resting on the felts and ready for the press was a

porse, or post. In Angoumois, a careful observer of the craft reported, a day's work consisted of twenty posts. That said, the contents of the posts actually varied enormously, once again depending on the dimensions and heft of the paper. For example, a post of *impériale*, a large, heavy paper that was difficult to work, included 79 sheets, but the same measure of *carré*, a standard printing paper, was composed of 173.[21] When the post reached the customary height it was pressed, the first of as many as ten or more times the paper went under the screw before it was ready for shipment. After the initial pressing, the layman began the delicate task of separating the damp sheets from the felts. He pinched the "good corner" and then inserted his hand carefully into the gap between felt and paper. When he reached the midway point he lifted the sheet firmly, freeing it from the felt. A careless layman could cause a great deal of damage, which is why Lalande considered this role "suitable only for people who have practised it from an early age and not for uneducated, inexperienced country-folk."[22]

Only thoroughly spoiled paper resurfaced as playing cards or cartons. Torn and crinkled sheets were marketed to less discriminating customers or hidden among the exquisite paper of quality reams. Long-cornered or serrated sheets were cut, and the best parts ended up as notepaper. The Montgolfiers recognized the value of their speckled and broken wares, exhorting every worker who touched inferior paper to treat it no differently than the good.[23] Meanwhile, female hands hung both sound and defective paper on cords to dry. The women moved the paper with a wooden, T-shaped frame, and inevitably marred their share. The toll was measured in scratches (*chaperons*) and leathery tracings called *maroquins*.[24] Moreover, the women or their supervisor had to monitor closely the movable slats of the drying loft. If the breezes robbed the paper of its moisture too quickly, it became brittle and failed to take the size (finish) well.

The sizerman (*saleran*) gathered the paper and immersed it in an emulsion of hides, hoofs, tripe, and alum. This gelatin bath filled the paper's pores, thereby preventing ink blots. The alum, the manufacturers said, enhanced the bond between the finish and the paper; the journeymen believed that it gave the sheets more "crackle." Certainly, size was an important expense in every papermaker's budget: it took one hundred pints of this brew to coat fifteen reams of ordinary printing paper, and every one hundred pounds of size contained five of alum.[25] Not surprisingly, the sizerman cooked his valuable mixture carefully, usually for thirty-six to forty-eight hours. If his fingers did not stick together when he dipped his hands into the pot, the size had the proper viscosity and was ready for use.[26] Still, the sizerman remained cautious and tucked only a small wad of sheets into his syrupy concoction. He then removed

one sheet from the dripping mass, dried it, and tested it with his tongue; if he left a balanced impression on the paper, akin to a fan or a butterfly's wing, and the sheet remained reasonably flexible, the size was good.[27] Size a shade too hot hardened the paper and ultimately flaked away. A mild, dry day was ideal for sizing. Too much humidity and the finish thinned and ran; on hot days, it dried too rapidly; in cold, it yellowed; and on stormy days, it turned. So important was favorable weather in this process that the sizerman's aides were the only female hands at Vidalon-le-Haut who were on call rather than expected at their benches every day.[28] (Taking no chances themselves, the Montgolfiers paid 1 livre 1 sou for an almanac in 1786.) The Montgolfiers demanded that the sizerman "always have samples of his [work] ready to show" and paid him well for mastering his capricious art.[29] Only the formaire earned more among Vidalon's workers.

The sizerman plunged "eight or nine quires [as many as 225 sheets] of small paper, or four or five [quires] of *grand raisin*" into his vat at once.[30] In order to distribute the finish evenly and squeeze out the excess (for rills marred the grain), the sizerman had the sticky paper pressed. Supervising the *colleuses,* the women who assisted in the sizing and hanging of the wet sheets, also absorbed much of the sizerman's time. He had to make sure that the women toiled with clean hands and completed one batch before moving to the next. The Montgolfiers also demanded that he prevent the colleuses and their children from eating in his shop or cooking in his vat, and to notify them if any of his charges quit.[31]

The colleuses separated the newly sized sheets with a shake or by gently blowing between them. They did so, wrote one admirer of their skills, "with a dexterity that surprises those who see it."[32] Like the layman, they always began at the "good corner" and always damaged numerous sheets. When one woman managed to detach half of a sheet from the pressed mass, her companion placed a wooden cross over its center. The first colleuse then folded the sheet over the cross, and her partner passed it onto the line. Two sheets draped over the cord together generally became inseparable waste, so the Montgolfiers pleaded tirelessly with the women of their sizing shop to pay less heed to their children and gossip.

After the sheets had dried for two or three days, Vidalon's *patrons* had them uncurled, stacked, pressed, shuffled, and pressed again. In smaller mills, virtually every worker tugged the bar at the great press before the screw was locked into place. Many manufacturers then burnished the glaze on both sides of their papers with a smoothstone; others attempted to hammer in a velvety shimmer. (But Nicolas Desmarest, the government's leading expert on the art, branded both techniques "false resources to mask the defects of

imperfect fabrication.")[33] Paper-sorters then scrutinized the finished sheets, removing dirt and excising sawtoothed margins. At Vidalon-le-Haut, the women separated the paper into five grades: good paper was white, intact, spotless, and of equal thickness; inferior was speckled, uneven, and hence inconsistently absorbent; *moitié* was cloudy, stained, wrinkled, and possessed highly irregular edges; *cassé*, or broken sheets, displayed tears and perforations, and could not be marketed as whole sheets; and *déchet* could only be returned to the pourrissoir for likely use in cardboard.[34]

The tellers scooped up the sorted sheets and blended them into quires. A ream consisted of twenty quires (five hundred sheets), but the balance in each turned on close negotiation between producer and consumer. For instance, one French fabricant offered to whiten his sheets with quicklime in order to capture the attention and purse of some Swiss printers. (Both parties doubtless knew that the printer's customers would claim years later that the letters in their handsome books had turned yellow.)[35] Bargaining was inevitably fierce, in large part because shaky hands, thick size, and drunken couchers meant that few, if any, reams were uniformly elegant or came near the government's standards for marketable wares.

The loftsman made up the reams, which Lalande maintained must consist of eight quires of good paper, eight more of inferior, and only four of broadly defective yet usable sheets.[36] After a final pressing, the new reams might be warehoused for a considerable time. As the proverb instructed, the best books were crafted with "ancien papier, nouvelle encre." Some storehouses may have bulged while veteran papermakers sought out reliable draymen, for small manufacturers could lose an entire season's revenue if a hauler disappeared with a shipment. Teamsters who could not find their way or left crates of paper in the rain tormented the Montgolfiers. Their business correspondence harped on the well-known shortcomings of these men, who probably became the scapegoats for production delays at Vidalon-le-Haut.

When the Montgolfiers' loftsmen needed ream wrappers, they turned to the stacks of spoiled sheets. About 6 percent of the paper produced in France, Desmarest estimated, was torn or broken, so the loftsmen had plenty of material from which to choose.[37] Even the sheets that were not ruined, however, bore traces of indifferent hands and imprecise judgments—judgments based on sticky fingers and the hairs of a fly's leg. Both paperworkers and their masters trusted in craft lore, experience, and a certain savvy to fashion a ream of paper. Theirs was a mechanical art, a series of delicate manipulations disciplined by hard-won, practical knowledge.[38]

To the Montgolfiers, this empirical wisdom was both a valuable heritage and a hurdle. To compete with the Dutch, they had to discard familiar prac-

tices and enter the realm of applied science, of calculated experiment. The Montgolfiers knew that to do so meant putting an end to the workers' capacity to maintain shopfloor routines and, even more, the occupational culture that sheltered them. In effect, Vidalon's *patrons* intended to make the journeymen's skills their own, revising and imparting them on their own terms. Here was the separation of conception and execution in the labor process that Harry Braverman spoke of in his classic work, *Labor and Monopoly Capital.* But the Montgolfiers reversed the poles of his case. Rather than "deskilling" by splitting complex skills into simple, repetitive tasks, Vidalon's *patrons* left the venerable division of labor intact.[39] Their ambition was to oust the journeymen's custom from their mill, thereby making skill (and the men who exercised it) subject to perpetual refinement.

Italian paper mill, date unknown.

Photograph courtesy of Robert C. Williams American Museum of Papermaking.

Sorting and cleaning the rags,
Encyclopédie, ou dictionnaire raisonné des sciences, des arts et des métiers.
Photograph courtesy of the Robert C. Williams American Museum of Papermaking.

French stamping mallets,
Art de faire le papier, 1761.
Photograph courtesy of the Robert C. Williams American Museum of Papermaking.

The tasks of papermaking,
Art de faire le papier, 1761.
Photograph courtesy of the Robert C. Williams American Museum of Papermaking.

The vat crew,
woodcut by Jost Amman, sixteenth century.
Photograph courtesy of the Robert C. Williams American Museum of Papermaking.

Sorting and checking the finished sheets,
Art de faire le papier, 1761.
Photograph courtesy of the Robert C. Williams American Museum of Papermaking.

CHAPTER THREE

The Montgolfiers and Their Craft

As every eighteenth-century publisher knew, paper consumed more than half of his budget. It represented 58 percent of the cost of producing Moréri's *Dictionnaire* in the early years of the century. Later, paper absorbed 55 percent of the expense of printing the third edition of Gibbon's *Decline and Fall of the Roman Empire*.[1] Not surprisingly, publishers and papermakers haggled passionately over the fine points of the weight and whiteness of the reams. And a third party's shadow, the readers', hung over this bargaining. The sort of men who purchased editions of Gibbon and Moréri during the Old Regime clothed them in Moroccan leather, accompanied by handsome endpapers. Such men cautiously perused the paper of the books they would add to their collections—books that were sold as masses of unbound, folded, cut, and carefully sewn sheets. Book buyers rubbed this paper between their fingers and hoisted it up to the light for a clear look at its knit, color, and blemishes. They shared an arcane vocabulary of quality, chattering about the paper's *force* and *oeil* (roughly, "luster"), and went into raptures about the azure tint of Dutch reams. Not every papermaker reached the readers' and hence the printers' standards; indeed, most turned out lackluster wares. The Montgolfiers, however, possessed a firm reputation as the producers of quality reams, although they, too, faced intense negotiations when their merchandise fell short of the mark.

Most studies of innovative systems of labor discipline, like those of Oberkampf, Wedgwood, Boulton and Watt, and Robert Owen, have centered on large shops and visible, strong-willed figures.[2] Many of these pioneers deftly transformed the technology or the organization of production in their fields; in fact, their efforts to police and train workers were often sparked by just such changes. Neither these men, their workshops, nor their ambitions were typical of their trades. The Montgolfiers fit the same mold. But their distinction should not obscure the problems they shared with papermakers large and small. From unpredictable weather to feckless teamsters, they were well versed in the conventional constraints of the craft. They employed the same basic technology and division of labor, endured the same competition from the

Dutch, and depended on the same headstrong journeymen as their lesser brothers. Yet, they also stood apart from all but a handful of French papermakers in more than the production of quality wares. They enjoyed advantages of location and climate, long production seasons and large scale, regular markets and webs of well-placed connections, substantial capital and a self-conscious patrimony of innovation. They would draw on all of these assets in their contest with the journeymen.

The Montgolfiers knew who they were. They considered themselves above the mere "speculators" who had scant learning or interest in their art, whose start-up costs had devoured their profits, and whose mills lay idle. "The vice of localism" did not hamstring the technological maturation of their shops, as it had in the mills established and continuously subsidized by provincial officials. Instead, the Montgolfiers identified themselves as entrepreneurs who owned and supervised their workshops, labored to advance the state of their craft, and as self-made men, had "abandoned mediocrity little by little." Their legacy was applied science, the "experiments they had made," and the desire to attempt more.[3] As Pierre Montgolfier explained, "His sons, who [were] eager to follow in his footsteps, [and] the manufacturers of the neighboring provinces who work[ed] for them and under their direction, [were] so many hotbeds from which the knowledge gained [of the art] will spread throughout their surroundings."[4] For David Landes, the intimate connections between the French firm and the family, whose fortunes, reputation, and style of life were tied to it, led to conservatism in all things. Such entrepreneurial dynasties were unlikely to seek outside capital or jettison familiar practices, even when more profitable and efficient production methods appeared.[5] But the Montgolfiers were not averse to cautious daring, especially when it provided them with the tools to challenge the Dutch.

Vidalon-le-Haut was located in the parish of Davézieux, a "twenty-minute" walk from the market and manufacturing town of Annonay.[6] A considerable village of about five thousand inhabitants in 1789, Annonay was perched at the northern tip of the region known as the Vivarais, itself a northern slice of the vast province of Languedoc. Its commercial currents were stirred by both nearby Lyons and its own agrarian cycles. An anonymous pamphlet from 1787 observed that "it is good for the people of Gévaudan, Velay, Vivarais, and Cévennes to take up manufacturing; but they have the sense to become manufacturers and farmers, depending on the season."[7] The people of Annonay cultivated silkworms, tanned skins, sheared sheep, wove and dyed woolens, softened leather for glovemaking, and sweated in a new ribbon mill.[8] They also knew something about labor strife: rival journeymen's associations battled in 1788 over access to employment in Annonay's textile industry.[9]

Annonay was built around the confluence of the Cance and Deûme Rivers. Vidalon's position on the Deûme provided the Montgolfiers with access to the fairs and markets of Lyons, as well as the opportunity to ship paper to Paris and Orléans. The Rhône opened the markets of Marseilles and the Mediterranean in general to them and, along with the Saône, brought the treasured rags of Burgundy. Desmarest reported that the sloping course of the Deûme was ideal for the "play" of millwheels and that old linen shredded in its clear, rushing water emerged surprisingly white.[10] But the Deûme also had its shortcomings, most notably its tendency to dry in the summer, silencing Vidalon's stamping mill for several weeks. Still, the Montgolfiers' access to largely regular water power distinguished Vidalon-le-Haut from a mill near Le Vigan in Languedoc, where Claude Chauvin supervised production campaigns of only six months "for want of water." A second enterprise close to Saint-Esprit "hardly work[ed] three months of the year" because the papermaker had to divert water to his grist- and fulling mills. While the Deûme supplied the fresh flows necessary for white, unstained reams, the fate of one manufacturer situated along the Tarn was encapsulated in the phrase "the waters are dirty most of the time."[11]

The Montgolfiers enjoyed a climate temperate enough to permit the sizing and drying of paper throughout most of the year. As Desmarest noted, this was no mean advantage.[12] At Uzès, well to the south of Annonay, the papermakers did all their sizing in the winter; in mountainous Tence and Saint-Didier-en-Velay, cold and humidity restricted the fabricants to sizing seasons of three months.[13] The Montgolfiers' physical resources, then, served as the base for their other advantages, especially their capacity to fashion large batches of quality reams.

According to legend, Jean de Montgolfier, a distant ancestor of the masters of Vidalon-le-Haut, learned the art of papermaking as a prisoner in Damascus during the Crusades. He transported this knowledge to France, founding both an industry and a family fortune. If nothing else, this tale echoed the Montgolfiers' sense that technological prowess was responsible for their elevated rank in the trade. As craftsmen, they shared information about workers, wages, and techniques with their fellow papermakers, or *"confrères,"* as they liked to call them.[14] They were the first masters among a cohort of masters and took this position seriously. About his "sacrifices . . . to establish order [*la règle*] and the subordination of the workers in [his] mill," Etienne Montgolfier boasted that "it was a policy [*planche*] that I undertook at my own expense and from which all of our *confrères* can profit."[15]

Yet, the Montgolfiers were also full of condescension for lesser producers bound by technological routine or languishing under the thumb of high-

handed journeymen. They wrote angrily to the millmasters who let workers evade the state's codes for securing new employment, especially when these men and women had fled Vidalon-le-Haut.[16] Simply put, the Montgolfiers knew their own interests and pursued them aggressively, in the belief that both their actions and vision enhanced the whole of the craft. "Nobody," Pierre Montgolfier intoned, "finds himself better situated to spread the usage of new machines and the knowledge of new procedures" through the trade.[17]

Perhaps the Montgolfiers' fairly rapid ascent from the ranks of the petty papermakers accounted for a certain swagger. In 1693, two Montgolfier brothers, Raymond and Michel, from Beaujeu in the Beaujolais, married the daughters of Antoine Chelles, then owner of Vidalon-le-Haut and its sister mill Vidalon-le-Bas.[18] Evidently, Raymond was an enterprising sort. He caught the tradewinds from Lyons and emphasized the production of large-format paper for business registers as well as wrapping for sugar loaves and the city's silks. In 1702, he placed the value of the two mills at 6,000 livres; by 1761, their joint worth was approximately 50,000 livres.[19] In 1735, the two Vidalons possessed five or six production vats, which was one measure of the scale of a paper mill; in 1812, for comparison, 82 percent of French *fabriques* housed only one vat, and these petty mills contained 62 percent of all French production units.[20] So Raymond was well on his way up the craft ladder when his oldest son, Pierre, inherited the larger mill, Vidalon-le-Haut, in 1743.

Marie-Hélène Reynaud playfully described Pierre Montgolfier as a seventeenth-century man conducting a nineteenth-century enterprise.[21] He was born in 1700 and had a brush with a religious vocation, but after studying at a seminary in Lyons he returned to the family business.[22] "Cold, regular, and punctual," Pierre rose at four every morning, regardless of the season, and washed his hands and face in the millrace. Every evening he retired at 7:00.[23] Stern and implacable, he was inflamed by both the journeymen's custom (and self-government) and the modest masters who knuckled under to their hands. Yet, he was genuinely open to technological change. If religious irreverence angered him, he was at home in the Enlightenment's embrace of applied science.[24]

Pierre fathered sixteen children. He favored Raymond, the oldest of his sons, with whom he formed a partnership in 1761 that included both Vidalons and the family's domains. (Antoine-François Montgolfier, Pierre's nephew, used his patrimonial rights to purchase Vidalon-le-Bas in 1770.) Raymond's premature death in 1772 thwarted his father's careful plan for succession at the primary mill, and complicated property and personal relations among the Montgolfiers.[25] When the aging Pierre turned to his youngest son, Etienne, to direct the affairs of Vidalon-le-Haut, the most the patriarch could offer was a

lease on the mill as a marriage portion. But Etienne's obligation was paid out of the mill's profits, and he alone enjoyed title to the new machines he installed there.[26]

According to Emeric David, a printer from Aix-en-Provence, Etienne Montgolfier was an affected man who conducted his business by the rules of mathematics. He had the aura of an academic, David continued, and was ruffled by haphazard experiment.[27] Etienne certainly counted himself among those "steeped in the science" of his craft.[28] The ethereal side of his nature rose only when he collaborated with his brother Joseph on their hot-air balloons—and in his passionate letters to his wife. Nevertheless, even this calculating man had to be reminded by his brother to turn the publicity surrounding balloon flights into coin by pursuing a state contract to supply the Indies Company with Montgolfier paper.[29]

Etienne had not been trained to succeed his father. Born in 1745, he had been sent to Paris to study architecture with the famous J.-G. Soufflot in preparation for a fitting bourgeois career as a mechanical engineer or architect. He was a *mondain* and a Freemason, the acquaintance of Condorcet, Franklin, Lavoisier, and Malesherbes, as well as the friend of the future Revolutionary, Boissy d'Anglas, and the noted wallpaper manufacturer, Réveillon.[30] At a more intimate level, he could count on his large family. His uncle Jacques, originally dispatched to Paris to establish himself as a wholesale paper merchant, had become the *Receveur Général* for the archdiocese of Paris. His brother Alexandre-Charles, a canon in the church, obtained a seat on the seneschal's court in Annonay, which served the family well in 1781, when he imprisoned several of Vidalon's striking workers.[31] And his turbulent brother Jean-Pierre (b. 1732), who had made his *tour de France* among the paperworkers and knew their ways, supervised the mill while Etienne promoted balloons and sought privileges in the capital.

For the most part, the Old Regime entrepreneur who looked beyond the bonds of kin for managers was thought to be at risk.[32] Pierre Montgolfier's considerable brood freed him from this concern. As he grew older, he delegated responsibilities in the mill, creating a division of labor at the top that was as precise and exacting as the regime imposed on his workers. Etienne was in charge of correspondence, oversaw the manufacturing process, and determined which papers were made. He governed the foremen and the formaires, experiments and new construction. Jean-Pierre kept track of output, the payroll, and the quantity of sorted and fermented rags. He was expected to purchase planks of wood and to alert workers about their shortcomings. His sister Marianne performed equally exhausting tasks. She monitored the rag-sorters and all the female hands who did not take their orders from the

sizerman or loftsman. She also supervised the *ménage*, which entailed responsibility for the flock of servants and valets exempt from the making of paper. Pierre himself assigned work, oversaw the drying rooms, kept the warehouse journal, watched over the finishing shop, and made sure that all tools were properly stored. Together with Marianne and Jean-Pierre, and perhaps to check his offspring, the old man weighed the rags, size, coal, and wood that Vidalon-le-Haut consumed.[33]

Of course, the petty proprietors of most mills needed no help managing the vat and its crew. Of thirty-eight mills recorded in a survey of the Languedocian craft in 1772, twenty-eight housed one production vat; seven others contained only two.[34] Elsewhere, the papermaker Jean Mourlhiou, a *cas typique* of the art, employed four wage-workers with no family connection to him, as well as his wife, three daughters, two sons-in-law, and two of their brothers.[35] In 1769, Vidalon-le-Haut alone housed and employed 113 adult workers, as well as 52 children.[36] Twelve years later, the "*groupe* Montgolfier" worked more than twenty vats (eight of which were at Vidalon-le-Haut) in nine locations.[37]

Recently, Colin Jones has spoken of the "ideological autonomy" of the Old Regime bourgeoisie.[38] All papermakers were surely capitalists, but the distinction between *les grands* and the small fry left little room for broadly shared assumptions. After all, Vidalon's *patrons* portrayed the "vast majority" of the kingdom's paper manufacturers as "only simple workmen" hamstrung by "blind routine, [and] incapable of improvement by themselves."[39] A Dauphinois observer put it more baldly: most of his province's master papermakers, even those who owned their mills, were "without means, without credit, likewise without talent."[40]

Whereas the Montgolfiers were free (taxes and tolls aside) to trade in the markets of their choice, many papermakers were mere tenants, obliged to sell their entire year's output to a particular printer or mercer. These batches, or *campagnes*, were often contracted for in advance of production, generally in exchange for the provision of precious, discarded linen. (Here was a system guaranteed to embolden journeymen to blend tough bargaining and the selective withdrawal of their labor.)

Ruining a season's production spelled disaster for the twenty-five papermakers (of thirty-eight) in Languedoc who had to answer to a hungry landlord.[41] Not surprisingly, this risk added to the technological conservatism of the craft and even led certain producers to surrender control over the final appearance of their wares. To avoid the tears and stains of sizing and finishing, vulnerable manufacturers entrusted these tasks to paper merchants and more skillful confrères. A fabricant near Clairac dispatched all but one

hundred of his annual output of four thousand reams to Marseilles for finishing, while a producer in Brissac had approximately two thousand reams smoothed and burnished at Beaucaire, Marseilles, Montpellier, and Nîmes, among others.[42] For the papermakers of Annonay, whose carriage trade and international markets turned on their reputation for fashioning "the handsomest paper in the kingdom," such loss of mastery over the grain and luster of their wares would not do.[43]

"The whiteness of their papers exceeds all others," wrote one official about the Montgolfiers and their crosstown rivals, the Johannots.[44] While most papermakers concentrated on cheap, gray wrapping and inexpensive, lightweight reams destined for printers, the Montgolfiers produced fine papers for artists, architects, composers, and engravers, as well as for ledgers and correspondence. Much of their printing paper rivaled the heavy white reams of *carré fin* used in the Neuchâtel quarto edition of the *Encyclopédie*.[45] Their merchandise was so consistent that it mocked Lalande's assertion that an acceptable ream must include no less than 40 percent of good paper.[46] During the year X (1801–2) of the Revolutionary calendar, the Montgolfiers classified 76 percent of their output as *bon*, 9 percent as *inférieur*, and 15 percent as broken or fit for pulping.[47]

As a result, Montgolfier paper found wide markets. Lyons, Paris and its hinterlands, and foreign markets absorbed approximately 60 percent of Vidalon's output in 1771. A decade later, just as the Montgolfiers were remodeling their shops along Dutch lines, Lyons and *la région parisienne* accounted for more than three quarters of the mill's production.[48] And in 1784, Vidalon's *patrons* contacted potential customers, sent samples, prepared orders for or shipped paper to clients in Annonay, Arles, Avignon, Bagnols-sur-Cèze, Bordeaux, Carcassonne, Charly, Cluny, Grenoble, Le Genast, Lunel, Lyons, Marseilles, Meaux, Montauban, Montpellier, Nancy, Nîmes, Nyons, Orléans, Paris, Perpignan, Pézenas, Ponsart, Pont-de-Vaux, Roanne, Saint-Esprit, Saint-Jean-en-Royans, Serrières, Strasbourg, Toulon, Toulouse, Tournon, Troyes, Versailles, Bruges, Constantinople, Geneva, Naples, and Turin.[49] For comparison, consider the case of two producers near Aubenas who transported their reams, like most papermakers in Languedoc, to local markets on the back of a mule.[50]

Jean-Pierre Montgolfier knew the place of his family's mill in the craft, but he wanted more. "The experiments and *mémoires*" of Desmarest had led him to dream: "I foresee," he delighted in 1780, "that we will be able to find new machines to simplify work, new procedures to perfect it, and finally reduce to calculation all the operations [of papermaking]."[51] His was the optimism of applied science, of methodical advance. Hadn't the papermakers of Annonay

"by a sustained, reasoned, and thorough study of their art carried it to the last degree of perfection"?[52] Here was the voice of that part of the Old Regime's ideologically autonomous bourgeoisie perched in the privileged "international circuits of designs and designers, colours and chemicals, styles and fashions."[53] The Montgolfiers, the Martin brothers (of chinoiserie fame), and Réveillon enjoyed the means to break free of the "blind routine" of their crafts. But the Montgolfiers still faced fierce competition from the most elegant paper of their day, the Dutch wares known as *pro patria*.

CHAPTER FOUR

Rags, Regulation, and Government Stimulation

Flush with the news that his family had won a government subsidy, Jean-Pierre Montgolfier wrote an expansive letter to his benefactors, "*nos seigneurs*" of the Estates of Languedoc. The Montgolfiers had secured a substantial grant of 18,000 livres to emulate Dutch practices and machines. "This project in which the province is prepared to cooperate," Jean-Pierre exulted, would permit his family to conduct the "experiments" necessary "to pull [papermaking] out of the routine to which it had been abandoned." As a measure of his thriftiness and "zeal" for the improvement of his art, he offered to provide the provincial regime with a detailed, annual accounting of both Vidalon's breakthroughs and failed tests. Rather than duplicate successful efforts, enterprising producers could devote scarce resources to unresolved problems.[1] Thus Jean-Pierre linked the state, applied science, and the bottom line as the mainsprings of technological improvement.

Economic historians are busily reassessing the ties between manufacturers and the French state, especially during the last decades of the Old Regime. No longer confident in a simple model of burdensome regulation and suffocating producers, these scholars properly emphasize a mixed pattern of enduring restriction, hesitant deregulation, and active encouragement.[2] While the state codified the weights and dimensions of more than fifty sorts of paper, it also sought to ensure that entrepreneurs enjoyed adequate supplies of rags and journeymen, and later, access to Dutch procedures. Its interventions were most successful when they reinforced desires and opportunities shaped by the market, such as the manufacturers' lust to turn out wares as fashionable as pro patria. But the state's will could neither overcome rising rag prices and the export of old linen, nor offset the masters' seasonal need for workers and acquiescence in their custom.

The making of paper in France evidently began during the middle of the fourteenth century, near Troyes. Two centuries earlier, a mill near Valencia was producing paper from flax and hemp pulped by grindstones. Easily torn and more fragile than parchment, these early sheets did not always find ready acceptance: in 1231, the Holy Roman Emperor Frederick II declared that all

official acts must be inscribed on parchment.[3] In time, this bias against paper vanished, particularly after the papermakers of thirteenth-century Fabriano (near Perugia) replaced the grindstones with calibrated mallets and the starch size favored by Spanish producers with gelatin. "From Fabriano paper goes to all the world" became the town's motto; equally, its refined papermaking technology quickly migrated north. By the middle of the fifteenth century, the French industry, with its centers in the Auvergne, Angoumois, and the suburbs of Paris, fashioned enough paper to meet the country's demand.[4] Yet, the golden age of Louis XIV was anything but bright for French papermakers. They lost foreign markets, particularly in England and Holland, and encountered increasing competition at home. Parliament banned the importation of a variety of French goods, including paper, in 1678, on the grounds that such consumption "exhausted the treasure of the realm."[5] After the repeal of this Act in 1685, English merchants resumed intermittent, but diminished, trade with France's paper suppliers. With war looming, the Dutch outlawed the import of French paper in 1671. Worse yet, the migration of skilled Huguenot journeymen prompted by the revocation of the Edict of Nantes (1685) robbed the French craft of much-needed talent and strengthened the papermakers' Dutch rivals.[6] By 1717, only 57 of 119 Auvergnat mills remained active.[7] And the reputation of Dutch paper ballooned so rapidly that French fabricants turned out sheets marked with the *"armes d'Amsterdam."*[8]

The relative decline of French papermaking mattered to Jean-Baptiste Colbert, the minister whose name has become synonymous with the strict regulation of manufacture. He estimated that paper ranked fourth or fifth in importance among French exports.[9] But Colbert could not turn to guilds for aid in restoring the craft, since papermaking remained outside the corporate structure of the Old Regime. Perhaps this status can be explained by the industry's maturation in the fifteenth and sixteenth centuries, long after the medieval establishment of many craft communities. Certainly, the regulation of papermaking never proved easy. To take advantage of the rush of mountain streams, many mills were located in isolated hamlets, far from the prying eyes of officials. Always capitalist, the industry was in the hands of men more sensitive to the reproaches of their customers than to those of the state. Still, Colbert launched a set of national regulations for the craft in 1671. His code was designed to halt the alleged deterioration in the quality of French paper and to prevent the labor turmoil that disrupted the flow of paper from France's mills. In fact, the decree merely affirmed everyday practices and hence had little impact.

Over time, the Colbertian legacy of precise regulations for papermaking—and sporadic enforcement at best—resulted in a spate of royal edicts. Broadly

speaking, they addressed three concerns: the irregular supply of rags, the insubordinate journeymen, and the production standards (and technology) that would assure sales. Always scarce, discarded linen had been the subject of a fifteenth-century complaint by Genoese papermakers, who believed they were "under the thumb [*tutelle*] of the rag merchants."[10] In 1772, the survey of Languedoc's industry reported that the shortage of rags had idled one of the four vats at a mill near Le Vigan "for a long time."[11] To remedy the problem, Versailles had prohibited the export of old linen in May 1697 and March 1727, under pain of fine and confiscation. Judging by later laments, these proclamations had no effect. By the middle of the century, the state had ceased policing the domestic rag trade.[12] Millmasters desperate to fill a campagne or make the rent were simply not about to defer to the niceties of royal codes. Accordingly, the price of old linen in France continued to rise, and rags were ferried to still more lucrative markets in Germany and Holland.

Discarded linen was the most expensive item in every papermaker's budget. At Whatman's mill in England, an enterprise of international repute, rags accounted for just under half of production costs in 1784–85, while labor, in this highly skilled trade, amounted to only 14 percent of expenses.[13] As Tables A and B demonstrate, the workers' food and wages constituted far less burdensome charges in Lalande's and the Montgolfiers' calculations than did cast-off clothing. Consequently, Vidalon's *patrons* haggled ceaselessly with a multitude of rag furnishers; in 1789 alone, they purchased old linen from 115 sources. Even this aggressive pursuit, however, failed to insulate them from the steepening rag prices that afflicted their craft. From 1773 to 1788, the price they paid for superfine linen, the base of their high-quality reams, rose at an average annual rate of 3.33 percent.[14] Small wonder that Vidalon's *patrons* wanted reliable men to work the precious, stinking pulp.

If any worker quit their mill without taking the steps "mandated by the laws of the kingdom," the Montgolfiers pledged to pursue him (or her). In 1785, they affirmed their intention "to have the king's ordinances observed to their full extent at [Vidalon-le-Haut]."[15] They knew, however, that theirs was a lonely policy. Few masters had the means to stand up to the journeymen, especially when surrounded by vats of perishing pulp. Equally, the "weakness of the other workers" prevented them from resisting the custom of the many.[16] Later discussions in this book explore the journeymen's self-government and indifference to Versailles' edicts. Here, suffice it to say that the state preferred prudence—inaction and occasional penalties—over force in the enforcement of its own codes for the proper conduct of paperworkers.

In 1688, 1727, 1730, and 1732, the French state attempted to reform papermaking with new decrees. Determined to leave its mark on every ream, Ver-

sailles set out exhaustive standards for tools, techniques, and labor relations in 1739. Two years later, precise prescriptions for the weight and measurements of fifty-eight kinds of paper fine-tuned the regulatory instrument.[17] Nicolas Desmarest, a strong proponent of industrial liberty, branded the exacting specifications of 1741 a "hydra." All that kept French papermakers from being devoured, he believed, were the "transgressions that the industry happily allowed itself." Faced with the alternatives of punishment or loss of their markets, the manufacturers had made a reasonable choice: they "had preferred to cross the law rather than their interests or those of their mill." Once the schedules disappeared, Desmarest contended, a "uniformity of conventions" would govern the commerce and production of paper. This private ordering of the craft "should dispense the government from fixing any other arrangement." Formats and watermarks would be determined by sustained use in the marketplace and conserved in "the steady relation" of consumer and manufacturer. Consequently, shifts in taste would not yield artifice and deception; instead, they would be tucked safely into the always slow "revolutions of carefully considered needs [*besoins réfléchis*]."[18] Moreover, as the Dutch had demonstrated, novelty in the design and appearance of products improved quality and captured markets.

Annonay's papermakers practiced Desmarest's convictions. "The length of each type [of paper] and the weight of each ream" they made, wrote an official, "[were] not fixed." Rather these producers "follow[ed] no rules other than the stipulations of those who gave them commissions."[19] As Etienne Montgolfier explained, "the manufacturer must conform to the buyer's ideas." "Regulations," he recognized, "do not shape taste."[20]

The papermaker who adhered to the letter of the law purified the water in his dipping and sizing vats by passing it through a sand filter. If he failed to do so, he faced a small fine. But the mill owner who risked the contamination of his pulp by refusing to enclose his pourrissoir chanced a penalty of 3,000 livres.[21] The decree of 1739 was laced with such specifications and promised a potential bonanza for the state in penalties. More important, however, was the outlawing of the Hollander beaters, the Dutch machine that pulped rags quickly and efficiently.[22] Nevertheless, experiments with these devices, including one by the Montgolfiers, took place soon after the ban, and some were funded in whole or in part by provincial governments.[23] Indeed, the Montgolfiers, who were quick to hold their journeymen to Versailles' decrees, candidly noted that "if [the papermakers] were scrupulous slaves to the disposition of this ordinance, which exclusively conserved the procedures in use at the time of its compilation, they would not have been able to take a step towards perfection."[24] Etienne Montgolfier breathed more freely because "the

part [of the edict] which concerns production [was] never carried out in full."[25]

The state relied on the papermakers themselves to carry out its regulatory mandates. The manufacturers of each "district" (whose extent would be determined by the intendants) were expected to elect wardens, or *gardes-jurés-visiteurs*, from their own ranks. These men were supposed to spy on their confrères, making sure that their reams and machines conformed to code, and to initiate legal proceedings against transgressors. Any merchant or manufacturer who refused to open his doors to these inspectors risked a fine of 500 livres. To sweeten the pot for hesitant wardens, the edict of 1739 stipulated that lightweight and undersized reams were to be confiscated, pierced through the middle, and reduced to stock, with half the value of the pulp destined for the inspectors' pockets.[26]

In practice, the gardes-jurés-visiteurs often encountered closed doors and worse. After a hotheaded producer threatened him with "twenty blows with a bar," one warden conceded, "I do not speak of the other manufacturers because the bad treatment I received at the first intimidated me so greatly that I did not dare to go [to their mills]."[27] For others, too, discretion was the better part of valor. In 1739, a papermaker in Bédarieux turned down the assignment on the pretext of being behind the times and hence incapable of correcting the flaws of his confrères.[28] In Languedoc, with its widely dispersed industry, the assemblies of manufacturers evidently met too infrequently to select inspectors. Wardens vanished forty years before the Revolution in the Auvergnat craft and probably never functioned at all in the papermaking hamlet of Chamalières.[29]

Still, some inspectors got through. A Thiernois producer received a fine of 100 livres and the added indignity of forfeiting eleven underweight reams.[30] The intendant of Languedoc heard from a warden in 1743 about *"papiers non conformes."*[31] Four Auvergnat fabricants drew penalties in 1741 because grease had been found on the tables in their finishing rooms.[32] But this regulatory animus was evidently spent by the end of the 1740s; uncollected fines and infrequent impoundments did not prevent manufacturers from catering to their customers' needs. For Desmarest, the real danger of the edict of 1741 was that this "ghost" might return as a reformed "scarecrow" and alarm the papermakers, particularly the innovators.[33] Meanwhile, Léonard Valade fils, a merchant papermaker from Mazamet, chafed that he had suffered as the only papermaker in town who had followed the regulations issued in 1739.[34]

In a bow to corporate structure, the inspectors were supposed to watch warily (and collect fees) while apprentices and journeymen negotiated the routine passages of their craft. To join the roll of master papermakers, the

aspirant had to present written confirmation of four years of consecutive service as an apprentice and four more as a journeyman, fashion a chef d'oeuvre in the presence of the local community of masters, and identify the qualities of paper placed before him.[35] Lacking the paternal authority of her husband, the papermaker's widow (*veuve*) could not engage new apprentices. She also had to inscribe the word *veuve* on the reams she manufactured; otherwise, she shared all the rights and privileges enjoyed by her late spouse.[36] Accordingly, the edict of 1739 advanced guildlike proprieties for the master's widow and upwardly mobile men. Most likely, however, these rites of passage did little more than celebrate successions occasioned by deaths or birthdays.

Perhaps the state's greatest imprint on papermaking had taken the form of taxation. The social commentator Legrand d'Aussy maintained that a duty on paper in 1680 had obliged 60 of perhaps 139 mills in Ambert and Thiers (Auvergne) to shut down.[37] Taxes that had been "sometimes suppressed, sometimes reestablished, diminished then augmented" distorted production cycles and ultimately increased the price of French reams.[38] Even the short-lived paper tax of 1748, which was lifted a year later, prompted several Rouennais manufacturers to suspend operations. The levy of 1771, although rescinded in part in 1772, was apparently a crushing burden. A curé defended Jean Ansedat, who was seized as a vagabond in Ambert in 1773, by citing "the severe economic difficulties of the town: the paper industry had been virtually closed down by 'the price of food and the tax on paper.'"[39] And, Annonay's manufacturers grumbled, internal customs took their toll on the commerce in paper: the duties at Valence and Lyons amounted to one fourteenth of the price of their merchandise.[40]

For the most part, however, the troubles of French papermaking in the eighteenth century cannot be placed at the state's doorstep. Watermarks slipped off molds, producers heeded consumers rather than the government, and even provincial regimes were complicit in violating Versailles' mandates. In 1763, the state faced the facts of practice: it permitted papermakers to employ whatever machines they wished. International competition, labor turmoil, technological inertia, and a conservatism born of fading glory had all contributed to the relative decline of the French industry. But all was not lost. Although the output of the mills of Ambert slipped from thirteen thousand quintaux of paper in 1738 to a low of about seven thousand in 1750, production ascended to nineteen thousand in 1776.[41] The number of millwheels turning in the paper mills of the "three valleys" of the Auvergne, the heart of the province's industry, increased 30 percent from 1750 to 1776.[42] Around Paris, several large enterprises appeared.

Still, the craft's revival was neither continuous nor universal, and was

ceaselessly haunted by Dutch competition. Lalande claimed in 1761 that no more than one hundred of the four hundred paper mills that dotted the landscape of Angoulême and Périgord "in the old days" remained active.[43] "Twenty-five years ago," he added, the mills of Franche-Comté "supplied a great deal of paper to Switzerland and to the Lyonnais, over and above the consumption of the province, but, for a few years now, the quality and the trade have declined, several mills are short of work and Switzerland is no longer obliged to seek its supplies there."[44] Versailles could do little to curb the rising price of discarded linen or undermine the journeymen's control of the labor markets. But it had the means and resolve to aid French papermakers eager to challenge the Dutch, particularly those manufacturers already rich in reputation, resources, and confidence.

It should have been the best of all possible worlds for France's papermakers. At the end of the Old Regime, mirrors, umbrellas, wallpaper, and writing tables, among other things, were within the reach of more of the French than ever before.[45] Book ownership gained ground, even among *le petit peuple*, and craftsmen substituted paper and *papier mâché* for wood in furniture, chinoiserie, and picture frames.[46] Too often, however, Dutch producers satisfied the French appetite for fine writing and drawing papers. And the Dutch competed ever more successfully in the international markets for quality printing paper, a slice of the trade that once bore a French imprint. Meanwhile, the inelasticity of the French industry was revealed in 1777, when the publication of the quarto edition of the *Encyclopédie* touched off unprecedented demand for heavy, white printing paper. Producers in France and adjacent Swiss cantons strained to fashion the 36 million sheets of fine paper consumed by this venture; even so, a paper famine in the winter of 1777–78 nearly compelled the Société typographique de Neuchâtel, the publishers of this edition, to shut down and lay off their pressmen.[47] Bureaucrats and printers alike realized that French papermaking needed the impulse of reform.

Of course, recognition was one thing, effective action another. For the flock of petty producers of cheap wrapping and writing paper for local distribution, Dutch competition mattered little. For the great Auvergnat manufacturers, with their substantial Parisian markets (especially for printing paper) and access to family and subcontracting webs when demand peaked, the Dutch were more of an inconvenience than a threat.[48] But for Annonay's manufacturers, the Montgolfiers and the Johannots, for whom the international trade in luxury writing paper was crucial, the Dutch posed a genuine danger—and remained an imposing yardstick. Thus their shaky claim: "The paper of Annonay is looked upon as the handsomest paper of the kingdom; we have even

frequently equaled the paper of Holland if we haven't surpassed it."[49] Nevertheless, while Annonay's giants whistled past the graveyard, Amsterdam's paper merchants demanded that the producers along the Veluwe suspend operations for one month in 1740, to reduce inventory and improve prices.[50]

Market forces alone, however, did not account for the reform of French papermaking, a process that rested on the emulation of Dutch devices and manual techniques. Concerned about the fate of a valuable industry, the state intervened with its agents, academics, purse, and privileges. And the diffusion of applied science as both an outlook and a method, both within the state's *bureaux* and the leading circles of producers, bolstered confidence that the Dutch were within reach.

At mid-century, most observers agreed that French paper was "very far from that state of perfection possessed by Holland for a long time."[51] To "approach the papers of Holland" and even "pass" for them in foreign markets was high praise for a manufacturer near Lille.[52] With its soft grain, velvet shimmer, and light-blue tint, Dutch paper was fashionable and competitively priced. For the most part, it was said, Dutch works were somewhat larger and better capitalized than their French counterparts. But the real Dutch advantages lay in their machines for the transformation of rags into pulp and their techniques for the preservation of the grain in smooth sheets. Lacking the force of mountain streams, Dutch producers depended on the wind to drive their stampers. Sea breezes, however, often failed to move the banks of hammers, and prolonged calm reduced vatfuls of rotting rags to waste. So Dutch producers worked with fresh linen and entrusted the shredding to a device perfected in the 1670s, the Hollander beater. An oval tub served as the frame for this machine. A metal or stone base, studded with knives, was fixed to the tub's floor; a horizontal cylinder, again armed with blades, rotated over the stationary bedplate. As wind or water turned the cylinder, the rags were drawn through the gauntlet of opposed metal and torn apart.

Dispensing with fermentation saved time and money: less linen was lost and more ended up as the substance of sheets. Two Hollander beaters did the work of eighty mallets and required less space, supervision, and water power.[53] These devices worked quickly, taking no more than one third of the time needed by the mallets.[54] The French fabricant who followed the entire Dutch procedure, Desmarest concluded, might cut his losses in the course of manufacture by 75 percent.[55] The firm uniformity of Dutch paper invigorated producers and entranced customers; simply put, the papermakers of Holland enjoyed considerable advantages of cost and quality.

The secret of the handsome surface of Dutch paper lay in a technique known as *échange*. Whereas French papermakers regulated the airflow in their

drying lofts to hasten evaporation, the Dutch conserved the dampness of their fresh sheets. They pressed their wares lightly, shuffling the packs of paper between each turn of the screw. Successive contact with new sheets flattened the rough spots in each and preserved a gentle grain. To match the luster of Dutch paper, the French beat their sheets with trip hammers and burnished them with smoothstones. Nevertheless, French paper retained a rough, uneven appearance, and its grain, so important in guiding the pen and determining the paper's worth, remained irregular. Small wonder that Dutch paper found ready markets in France. Its elegant appearance, reduced production costs, and lively sales promised big profits—as much as 3.6 times the return of routine French methods, according to one advocate of the Dutch system.[56]

By 1750, certain French officials and papermakers were aware of the benefits of the outlawed beaters. As early as 1734 and 1736, descriptions of the devices had been published in Amsterdam, but installing effective replicas in France proved difficult.[57] In one attempt, the provincial government of Burgundy had dispensed 50,000 livres "in vain."[58] In 1751, the Montgolfiers and Johannots jointly drafted some Swiss workers to mount Hollander beaters in their mills in Annonay. But the machines were too heavy, required too much water, and yielded sheets that lacked "tenacity."[59] The want of journeymen capable of repairing the devices also burdened the papermakers, so they decided "to sacrifice [their] large outlays" and return to stamping hammers.[60] Dutch pro patria, however, continued to menace their markets, and Antoine-François Montgolfier (of Vidalon-le-Bas) marveled that the manufacturer equipped with a brace of beaters possessed "the ability to double the work of his mill."[61] Accordingly, Pierre Montgolfier petitioned Versailles for funds to sustain one of his sons in Holland while he gathered precise information about the Dutch art and its machines. He addressed his proposal to André-Timothée-Isaac de Bacalan, a commissioner in the Bureau of Commerce.

Pierre had reason to expect a sympathetic hearing from Bacalan. In 1768, the year before Pierre's request, Bacalan gave outspoken form to the cry "laissez-faire, laissez-passer": "Liberty is without doubt preferable to any rules. . . . All rules are absurd and injurious."[62] As the 1740s drew to a close, only embers still flickered from the state's comprehensive regulations of 1739. Yet, this unwieldy edict evidently had one lasting, ironic effect. As Pierre Claude Reynard has suggested, it led to a more active dialogue between papermakers and Versailles. At first, the manufacturers vented their opposition to a presumed *"carcan administratif."* The state soon retreated, both in the letter and the enforcement of the law. Its policies evolved in tune with the voice of the papermakers themselves, particularly large figures like the Montgolfiers, Johannots, and Réveillon.[63] Rather than stew in a Tocquevillian silence of

"submission and dependence," such entrepreneurs were quick to make their will known. Equally, they did not accommodate to an overbearing state's pressure to "adopt the best methods."[64] Instead, these innovative grandees and even like-minded small fry pressed the government to provide them with the means—and technical guidance—necessary to improve their reams.[65] Here the state and the market overlapped, with results far more successful than the regulation of the rag trade or journeymen's wandering.

The Bureau of Commerce was the government agency that supervised industrial activity in Enlightenment France. This office received its title in 1722 and a revised organizational scheme in 1724.[66] At its summit were the four intendants of commerce. (These men should not be confused with the provincial intendants, each of whom administered a *généralité*.) In the daily business of mapping a course for French industry, the intendants of commerce had their differences about the proper balance among regulation, instruction, and the judgment of markets. These tensions often filtered down to the inspectors of manufactures, the men on the ground who carried out the Bureau's mandates.

At first, the Corps of Manufacturing Inspectors, established in 1669, policed the work of the textile guild wardens.[67] In 1785, Jean-Marie Roland de la Platière, political economist and future Revolutionary, counted forty-two inspectors, whose responsibilities ranged across French industry.[68] Each of these officials was assigned to a généralité, but several, like Nicolas Desmarest, who had immersed himself in the art of papermaking, acquired a particular expertise that led them through several provinces. Most inspectors remained loyal to their initial charge, the strict enforcement of the royal edicts that governed manufacturers and their products, or endorsed a middle way of moderate regulation. Only a handful shared Desmarest's willingness to accept the market's verdict.[69]

Desmarest had the makings of a fine inspector of manufactures, said the future Controller General and economic reformer Anne-Robert-Jacques Turgot. He trusted in observation and experiment, hands-on experience rather than the dry voice of distant authority, and an accessible, regular order of nature that could be turned to productive use. He was a budding scientist, whose study of the volcanic origins of the basalt columns of the Auvergne landed him a place in the Royal Academy of Sciences in 1771. That Desmarest explored the basalt prisms while acting as an emissary of the Bureau of Commerce reflected his single-minded approach to problems of every sort.[70] Turgot delighted that the young scientist preferred "the route of instruction" to the "rigorous habit of seizures and fines too much used by his predecessors."[71] Certainly, Turgot needed a teacher more than a reflexive regulator in

his rocky, backward intendancy of Limousin. So he lobbied for Desmarest's appointment as inspector of manufactures in his généralité. Desmarest obtained this post in 1762, with express authority to probe the affairs of the dyers, tanners, cloth and woolens producers, and papermakers. Few officials of any kind knew enough to assess or guide papermaking. Most shared, if not as candidly, the thin understanding of the craft voiced by an administrator in Rennes: "Never having been inclined to study the fabrication of paper in depth . . . I have sought instruction more than brought the capacity to instruct."[72] Desmarest also turned to the papermakers for enlightenment, with an eye toward reforming the entire craft.

Desmarest believed that the Dutch method of papermaking brought together a systematic approach to technological improvement, relentless attention to the appearance of the reams, and careful consideration of fashion and competitiveness, all animated by perfect liberty.[73] It was this ideal combination of setting and practice, he claimed, that had given the Dutch their edge. He condemned the edict of 1741 as a misguided effort to freeze matters that "should naturally be subjected to the caprices of fashion and the needs it breeds." He mocked the government's apparent conclusion in 1739 that "the art is perfect." Rather than the "imaginary order" of these ordinances, he trusted his experience—experience that led him to reject both the state's prescriptions and the likelihood of the successful, natural migration of Hollanders and échange to French soil.[74] Hamstrung by prejudice and superficial borrowings, French papermakers had failed to master Dutch practices. These defeats, Desmarest knew, had prepared the ground for proper state action: the government would secure and circulate precise knowledge about Dutch papermaking and convince the manufacturers of the virtues of their rival's art. To penetrate the mysteries of pro patria, Desmarest petitioned the Bureau of Commerce to send him (rather than one of Pierre Montgolfier's sons) to the Zaan region of Holland. He made the trip in 1768.

Journeying to foreign parts in pursuit of an advanced tool or technique was common practice in papermaking. So, too, were attempts to entice strangers to abscond with innovations hidden in their kit. When a sixteenth-century English stationer attempted to lure French paperworkers across the Channel, his plan was frustrated "by practisinge the destruccion of the workmen and by writing and calling them Traytours to their Countrey, and sending men of purpose to Slaye them, as it hathe byn credeably declared unto me."[75] As the pace of technological change quickened in the eighteenth century, more conventional means were used to prevent unwanted transfer. For example, in 1786, England prohibited the export of papermaking equipment.[76] Such measures had little chance of success. After all, the instruments of papermaking,

including the Hollander beaters, were simple. But the skillful handling of these devices proved as difficult as mastery of the allied techniques that transformed Dutch practices into a distinctive craft. What distinguished Desmarest's voyage, then, was his desire to acquire the knack, to obtain a full sense of the Dutch art.

Desmarest was one of many technological informants traveling about Europe in the middle of the eighteenth century. (Perhaps Gabriel Jars' voyage through Britain, Sweden, and central Europe in search of advanced approaches in metallurgy is the best known journey of this sort.)[77] These missions were both private and accomplished with the connivance and purse of the state, both furtive and conducted in plain sight with Academy passports and aristocratic hauteur.[78] Certainly, pressure was mounting in French administrative circles for an informed inspection of the Dutch art, despite the failure of two, earlier state-sponsored treks.[79] For his part, Desmarest toured Broek, Gelderland, and Leiden but concentrated on the mills along the Zaan, which he explored "in the greatest detail."[80] On his way home he stopped in Flanders, where he examined Hollander beaters powered by water rather than wind, a technology he intended to test in France.

Desmarest presented his findings to the Royal Academy of Sciences in 1771 and 1774. His first mémoire (report), which appeared in the Academy's proceedings in 1774, celebrated the careful drying and pressing techniques employed by the Dutch. His second mémoire, published in the same proceedings in 1778, centered on the advantages of working with fresh, nonfermented rags and Hollander beaters.[81] Doubtless, he intended these tracts to possess the disinterested voice of science, but they were rich in advocacy and self-congratulation. "We know how slowly," Desmarest wrote, "the new practices of the most useful arts spread from one state to another when no one is charged expressly to follow or to hasten their progress." Before his voyage to Holland, "no *artiste* truly immersed in the Dutch procedures" had participated in the effort to emulate them.[82] Why was the vast enterprise at Montargis, the paper mill depicted in the *Encyclopédie*, hobbled by beaters so debilitated that they could shred only fermented rags?[83] No one in France, Desmarest explained, had engaged "in precise research on the play and supervision of these machines, on the details of their work, and finally on the principles of a good trituration [the maceration of the rags]."[84] France's papermakers had failed to visualize the whole Dutch system. The convictions and sheer "prejudices" they had distilled from their own routines blinded them.[85] Thus Desmarest scoffed at his countrymen's claims that the Dutch calendered their papers, drew on a superior labor force, or worked their men slowly. To those who located the Dutch advantage in choice rags, he sug-

gested a quick experiment with conventional French methods and the finest linen. It was the integrated, considered Dutch methods—the use of fresh rags, slow drying, échange with newly minted and sized sheets, costlier felts, and the beaters—that mattered. Desmarest had brought the systematic knowledge of Dutch papermaking home.

Dutch reams, however, had their flaws. It was said that the paper of the Zaan region was more liable to tear than French sheets, more resistant to the imprint of copperplates, and more likely to be pierced by new printer's type. Supposedly, Dutch paper broke more easily when folded, a considerable liability in an age when books were generally sold as masses of folded, sewn, and sometimes cut sheets. Desmarest patiently explained that the qualities that rendered Dutch paper so valuable for writing, drawing, and wrapping naturally undercut its use at the press.[86]

This admission drew Desmarest into a discussion of enlightened practice of the craft. "It was dangerous and false," he cautioned, to give familiar paths "an indefinite extension." But he did not wish to suppress French methods, which had their virtues for the production of printing paper and playing cards. He had "nothing more at heart" than to make French practice "more sure and more regular." In fact, Desmarest called on manufacturers to master the fine points and inner connections of both approaches. He envisioned a new kind of papermaker, who made informed choices between technologies and positioned himself "to satisfy the demands of every customer."[87] Long before historians discovered "flexible specialization," Desmarest championed it. In doing so, he reversed mercantilist assumptions about sales. Attending to shifts in taste and possessing studied, flexible technologies were the keys to prevailing in the market. Rationally determined products would underscore the rational play of markets.

But all that was in the future. First, Desmarest's discoveries had to reach a broad audience of papermakers. Fortunately, his case found favor with his superiors, and copies of his treatises, printed at royal expense, soon circulated across France. The reports were funneled through the provincial intendants, who inevitably asked for more than their original allotments. The intendant in Alençon pledged to exhort the papermakers and inspector in his charge to study the first mémoire, in exchange for six more copies. To disseminate the Dutch procedures in the Gironde, the intendant in Bordeaux needed an additional two dozen. The skeptical intendant in Rouen planned to circulate the twenty-five copies he received but expected slight returns. He agreed "to urge [the papermakers] to profit, to the extent it is in them," from Desmarest's views.[88]

Diffusing Desmarest's mémoires was wise policy, Antoine-Laurent de

Lavoisier and Jean-Baptiste Jumelin argued. It was also the sum of appropriate state intervention in the reform of industry: "When the government has spread instruction and knowledge [*lumières*], it has done all that it can do, all that it has a right to do." Then it must stand aside and rely on "this always active force" to work its magic. "In a word," this impulse was "self-interest."[89] But interest in French paper made with Hollander beaters lagged, an indifference that Desmarest traced to imprecise machines and the manufacturers' reluctance to adopt the complete Dutch system. French-built beaters clogged easily and frequent contact with the bedplate dulled their blades, producing "papers as defective as the papers made according to the usual method."[90] Even the director of the mill that produced some of the best faux-Dutch wares conceded, "There are many manufacturers in France who regard the imitation of Dutch paper as impossible."[91] Desmarest and the Bureau of Commerce needed a successful, homegrown model of the Dutch art, a three-dimensional nudge. For that, he required more precise information about the design and workings of the beaters. Even before the second mémoire appeared in print, Desmarest knew he had to return to Holland.

In 1777, Desmarest left for Holland in the company of Jean-Guillaume Ecrevisse, a Flemish artisan-draftsman. The son of a papermaker, Ecrevisse certainly knew his way around a paper mill. He had toiled as a paperworker, mill carpenter, and millmaster, studied the technical journals, and mounted a brace of beaters in a fabrique near Lille. In Holland with Desmarest, he posed questions in Dutch and scribbled away; consequently, the two travelers returned to France with "an ample harvest of documents," not the least of which were Ecrevisse's sketches of the Dutch machines.[92]

Desmarest's campaign for a pilot mill, equipped with efficient beaters and a shop devoted to échange, began in 1774 and gained momentum on his return from Holland. The enterprise would be open to all comers and demonstrate "the order and connection of the operations, their sequence and progression." Producers could observe devices and techniques "tested and verified by results," and would be free to duplicate them.[93] This was papermaking as applied science, with close regard for both rationalized products and the ledger. When an early proposal for state funds to establish this enlightened enterprise in Angoumois failed, Desmarest turned toward Annonay.

He was especially pleased that both the Montgolfiers and Johannots, driven by their "zeal . . . for the perfection of their papers," had mounted workshops for échange on their own, despite the high start-up costs.[94] He found a kindred spirit in Etienne Montgolfier, whose passion for carefully constructed experiment matched his own. He celebrated the Montgolfiers' inventive method for extracting gelatin from slaughterhouse remnants, which

produced a "very clear" size.[95] He joined Etienne in a series of tests that combined traditional French stampers with fresh rags. After one of these trials, the manufacturer explained to the inspector that his unfermented filaments soaked up too much water and lacked the "lightness" that Desmarest had attributed to Dutch pulp.[96] To produce commercially competitive wares, the manufacturers of Annonay needed Hollander beaters.[97]

Installing these machines in Annonay was an expensive proposition, as the beaters were not prêt-à-porter. They had to be shaped with local conditions in mind, including the elaborate system of races and canals that offset seasonal shortfalls of water power in Annonay. New men would have to be trained to govern the beaters; new, deeper molds—another distinctive aspect of the Dutch process—would have to be purchased. Desmarest put the cost of transforming a French mill along Dutch lines at 18,000 to 20,000 livres, a considerable sum.[98] He turned to the Estates of Languedoc, then better fixed to provide capital than Paris, to defray the expense. There would be a competition for the funds, but it was most likely rigged in the Montgolfiers' favor; indeed, both Mathieu Johannot and Antoine-François Montgolfier complained they never knew that the subsidy was available.[99] In 1780, the Estates of Languedoc adopted the project; a year later, it pledged 18,000 livres to the Montgolfiers and their mill, Vidalon-le-Haut. Three years later, they were still trying to collect half the money.[100]

In exchange for their evanescent subsidy, the Montgolfiers were to (1) mount two Hollander beaters within eighteen months, one to macerate the rags and the second to pulp the filaments; (2) equip their shops with all the tools necessary to imitate the Dutch; (3) follow each step and procedure of the Dutch process; (4) fashion every sort of paper that had contributed to the reputation of Dutch wares; (5) pay Ecrevisse, whom Desmarest had already dispatched to Annonay to oversee the construction of the beaters and the instruction of a flock of workers; and (6) open their workshops to their fellow manufacturers.[101] Vidalon's *patrons* kept their part of the bargain, permitting manufacturers from their province and the kingdom at large to observe "all the operations" and "follow the construction" of the beaters.[102]

Desmarest and Ecrevisse soon moved on to Essonnes, near Paris. Frustrated by an "ignorant" manager, the owners of the large mill there implored Desmarest to place their enterprise "on the same footing as that of Annonay." In fact, this fabrique also took on the contours of a pilot mill, since it housed the first Dutch-style drying loft in France. Word of the advantages of this "beautiful, rational edifice" spread swiftly: Desmarest boasted that papermakers from "a large number of other mills" came by "to take [the loft's] measurements in order to build one in their own mills."[103] At Annonay, he

explained, and apparently at Essonnes as well, manufacturers could assess the tools and tasks "which would best suit their ideas or business."[104]

This informed choice, Desmarest imagined, would loosen the grip of custom and the merely practical. Producers would overthrow the rule of thumb in favor of the precise knowledge he had made available. In the hardscrabble Angoumois, the site of Desmarest's early assignment with Turgot, he heard an echo of this design. In 1784, Desmarest helped secure funds for the conversion of Henri Villarmain's fabrique at La Couronne into a model Dutch mill.[105] Villarmain was well placed in the politics of French papermaking and, no doubt, tuned his pursuit of assistance to please Desmarest's ear. That said, Villarmain made the link between state intervention and a scientific approach to the reform of his trade. He knew of the early, "fruitless" efforts to emulate the Dutch at Montargis, in Burgundy, in Lorraine, and in several mills in Angoumois. "*Les lumières*," he claimed, "were missing then." Now, "thanks to the pains of the government," papermakers enjoyed "a body of doctrine of sure principles, a precise theory of the connected procedures [of their art]." Desmarest's mémoires had offered the industry reliable footing: "Discarding routine," Villarmain's craft "ha[d] become a rational method, an integrated process."[106] Simply put, the papermakers encountered a state committed to their technological advance and complicit in their evasion of its own antiquated regulations.

In 1783, Pierre Montgolfier was ennobled.[107] A year later, his family's mill became a *manufacture royale* (royal manufactory), enabling him to mount the king's insignia above the mill's doors. In part, Vidalon-le-Haut received this honor as recognition for the Montgolfiers' role in a technical advance, the introduction of "wove paper" in France. A fine brass screen, laced together on a loom, replaced the wires of a traditional mold in the making of wove paper. Rather than sharp impressions, the threaded brass left indistinct traces and sheets of a more uniform thickness.[108] This process enhanced the quality of Montgolfier paper; the royal escutcheon confirmed its worth and promoted its virtues in the market. Accordingly, the state and Vidalon's *patrons* had entered into a sort of partnership, an alliance rooted in technological progress.

This connection was anything but a Tocquevillian imposition. It did not represent a meddlesome government forcing best practice on reluctant or ill-prepared producers. Instead, it reflected a newfound reciprocity between technically adept agents of the state and progressive producers. Applied science was the relationship's bond, the entrepreneur's elevated standing and reputation for technical virtuosity its groundwork. Desmarest hoped that its success would invigorate "a powerful and public *motif* of emulation" in the

craft.[109] Meanwhile, the inspector and Vidalon's *patrons* were sure of one thing: the need to match the Dutch reliance on advanced machines, especially as the replacement for skilled hands. French papermakers, the Montgolfiers thundered, must make "fabrication as little dependent as possible on the pains of workers on whom one can rarely count, perfect the machines which are employed in papermaking, and imitate the Dutch who substitute them for men as much as possible."[110] Desmarest marveled that "the Dutch, while diminishing the work of men, are committed to extract the greatest services from their machines; it is from these perspectives that they construct and maintain them with attentions and outlays which always astonish."[111]

Surely, the French state's advocacy and purse extended its reach over the papermakers' technology far more than had reams of regulations. Yet, the government's mandates rested on more than the compliance of the fabricants: they required their active cooperation. When private and state interests dovetailed, such as in the diffusion of the beaters, the thump of the machine was soon heard in fourteen généralités.[112] But in such matters as the supply of discarded linen and the control of the journeymen's tramping, where seasonal markets trumped Versailles' ordinances, the influence of the Council of State ended at its doorway. Rapacious entrepreneurs and aggressive journeymen had long established their own customs, incentives, and penalties to police the trade. Thus the building of Hollander beaters and, even more, the Montgolfiers' training of a swarm of newcomers innocent of the journeymen's ways, inevitably sparked conflict.

PART TWO

The "Modes" and the Lockout of 1781

CHAPTER FIVE

Building the Beaters and the Journeymen's Custom

Jean-Guillaume Ecrevisse arrived in Annonay in May 1780. He had been dispatched by Desmarest to construct Dutch shops at Vidalon-le-Haut and was sure he could do so. A decade earlier, he had transformed a lesser mill near Lille along Dutch lines, and its product passed for pro patria in the market. This experience taught him that the "perfection" of the art was a matter of design and machines rather than the discernment of skilled hands. He scoffed at those manufacturers who attributed the defects of French wares to the flaws of French workmen. "We have no workers like the Dutch," they lamented. But their concern was misplaced, Ecrevisse countered, because the pulp determined the suppleness, the whiteness, and the unity of their paper, as well as its capacity "to drink the size." "If the hand of the vatman could produce these qualities," he continued, "nothing would be as easy as that improvement." It would require "neither construction, nor expense, nor knowledge."[1] Still, Ecrevisse was well aware of the value of skill in papermaking and the need to reward it. He would only consider the directorship of a fabrique near Amiens, he wrote in 1779, if he enjoyed "absolute authority, without restrictions, to hire and fire the subjects or workers as it pleases me." He also demanded the right to set the journeymen's wages "in proportion to their capacity."[2]

As historians reassessed the nature of proletarianization, they inevitably altered their understanding of skill, especially before the onset of large-scale mechanization. Rather than restrict proletarianization to the replacement of skilled hands by machines, the concept was broadened to include the loss of mastery over the work process. Subcontracting, the fracturing of skill in widened divisions of labor, and a host of craft-specific threats to the artisan's independence were all recognized as motors of proletarianization.[3]

In his influential study, *Work and Wages*, Michael Sonenscher took this line of thinking one step further. In place of distinctive tasks and closely guarded secrets, he emphasized the common skills and raw materials employed by the craftsmen of Old Regime Paris. Moreover, these men had been integrated into production networks, webs of artisans from different craft communities jointly working up and refining a particular product. These links naturally undercut

stable divisions between the trades and diluted identities. To ensure their status and wages, the craftsmen resorted to cultural artifice: they invented rituals, customs, and celebrations that enhanced their standing and distinctiveness, since the uniqueness of their daily tasks no longer did so.[4] Perhaps this was the case in some trades, but papermaking presents a different picture. The workers' skills were unlike those of the corporate crafts, as were the materials and instruments with which they toiled. Furthermore, their custom both reflected and lionized the particularities of their trade. They needed no artifice to remind them that they were "of another essence, different and more distinguished" than other journeymen.[5] Their skill had rendered them so.

Both the paperworkers' skill and custom were exercised in the market. Indeed, Sonenscher rightly questions the notion that craft custom separated and insulated the skilled from the tumult of the market during the twilight of the Old Regime.[6] Instead, the confrontation at Vidalon-le-Haut in 1781 pitted the journeymen's market-based custom against the applied science of Desmarest, Ecrevisse, and the Montgolfiers. Put another way, the state technicians' and the entrepreneurs' search for unfettered space, in which they could manipulate technique freely, placed them on a collision course with the workers' custom and the skills that undergirded it. Thus the Montgolfiers' beleaguered depiction of a *successful* experiment with paper dyes: "This test succeeded despite the opposition of the workers [who are] always difficult to lead when one removes them from their routine, which is not one of the least shackles to the progress of the perfection [of papermaking]."[7] Small wonder that the Montgolfiers chose to follow Desmarest's counsel and "instruct youths in the new procedures" rather than retrain their old hands.[8] If French papermaking was regulated effectively before the introduction of the Hollanders, it was from below, by the journeymen. In locking out their veteran hands in 1781, the Montgolfiers pursued the liberty to engineer one corner of the craft from above. Dutch competition had prodded them to do so; Desmarest's and Ecrevisse's expertise made the attempt to deregulate their shops worthwhile.

The usual "inconveniences which accompany a new project" had blocked the Montgolfiers' adoption of Hollander beaters during the early 1750s.[9] Among these, Joseph and Augustin Montgolfier claimed, were the efforts by their father's "workers [that] forced him to return to conventional practices," that is, stamping hammers.[10] Not surprisingly, tension surrounded the Montgolfiers' construction of their Dutch-style shops in 1780–81. The building that housed the three beaters, four new vats, and presses, was 22.5 meters long and 12 wide. It was fed by a canal of hewn stone that was 58.5 meters long, which finished in a decanting basin.[11] Evidently, the Montgolfiers had mounted two

working beaters and three vats by the end of 1780, well ahead of the eighteen months mandated by the Estates of Languedoc.

They divided production along the lines advocated by Desmarest, devoting their new mill to fine and superfine writing paper, and to reams destined for drawing, sheathing, and wrapping. They also supplied the Royal Sugar Refinery in Montpellier with the violet papers used in the transport of cones of sugar, thereby ousting the Dutch from a lucrative market.[12] And they hiked the prices they charged for their reams, already the highest in France.[13] But their good fortune was not without its troubles. In November 1781, Etienne Montgolfier lamented that he could not leave his family's mill for a moment. Labor problems, in the form of "routine-minded" journeymen who would not "abandon their old habits," were tormenting Vidalon's *patrons*. To counter "an insubordination [that was the] enemy of any improvement," they were schooling "young pupils in the Dutch procedures."[14] But the rapid growth of the mill had left the Montgolfiers vulnerable, and the journeymen, pressing ever harder, knew it.

In 1769, Pierre Montgolfier damned the journeymen paperworkers' "self-styled laws," or *"modes,"* as they called them. He maintained that these customs and conventions "lead only to disorder," but he also realized that humbling cocksure journeymen was easier said than done.[15] Time and again, master papermakers and bureaucrats alike reviled the spirit of independence and esprit de corps of the workers. It was as if this "mob" (*engeance*) had established a separate polity, a realm in which their writ alone was legitimate.[16] The proper order of the craft had been inverted, said ministers and manufacturers; the "first rule" of France's paperworkers was "to be the despotic masters of their bosses."[17] "Nothing is more revolting," Pierre Montgolfier intoned, "than the tyrannical power which the worker wields with respect to his master."[18] "The master [papermakers] are like slaves of the journeymen and workers," sputtered an anonymous mémoire.[19] In 1771, one observer cast all this in political terms: "The journeymen paperworkers form a sort of little republican state in the midst of the monarchy."[20] The king's men were concerned too: writing in 1772, a provincial official in the Massif Central worried about "a republic of inferior workers, accustomed to laying down the law to the masters."[21] He knew that journeymen paperworkers had forged local, regional, perhaps even national associations that defined and expressed *their* social responsibilities and monitored *their* acts of mutual assistance. Here were the journeymen's own civic bodies, nagging reminders that more than one civic tradition persisted in the trade. Master papermakers and journeymen paperworkers did not consider themselves to be members of a unified craft

community with a single set of interests. Rather, the bosses charged that the workers were willing to push them into bankruptcy or shutter a mill in defense of the "modes."[22] Consequently, paper producers and their skilled hands struggled ceaselessly to impose their separate versions of bon ordre on the entire craft.

Early in the nineteenth century, a petty papermaker explained why the "modes" outlasted the French Revolution. The "causes which favor [the journeymen's] abuses and oblige the masters to be mute" were many. Above all, this Auvergnat producer maintained, was the risk that the journeymen would withdraw their labor. Even the departure of just the vatman or coucher shut down production, thereby threatening the loss of a customer or failure to meet the rent. A prolonged suspension of work also meant that the linen fermenting in the pourrissoir would be "entirely spoiled." Moreover, the papermaker who hauled "a mutinous and insubordinate worker" into court was menaced with the torching of his mill, a threat that was enhanced by the journeymen's custom of commencing work around two in the morning. Fines levied by the workers against recalcitrant bosses or wayward shopmates quickly brought both into line, while swelling the treasury of the journeymen's combination. The willingness of millmasters to hire tramping men without proper papers only served to fuel the journeymen's license, as did the "negligence" of the authorities in chasing after the undocumented. Finally, the absence of gardes-jurés-visiteurs left the wronged manufacturer in the uncomfortable, face-to-face position of pursuing a shopful of malefactors on his own. And so the papermakers submitted to their workers' "laws."[23] The result, concluded an observer of the Dauphinois trade, was "an abusive routine which tends to the ruin of the business."[24]

Skill was the cornerstone of the paperworkers' powers and autonomy. This mastery enabled them to set terms for the appropriate exercise of their art and its transmission across generations, and to thin their ranks. Ironically, the paperworkers' numbers, especially in industry centers like Ambert and Thiers, also enhanced their might. "This is perhaps not the hour to take a rigorous approach" with these journeymen, wrote one official.[25] "The workers," an inspector of manufactures claimed in 1778, were "too numerous and too united for a brigade of the *maréchaussée* [mounted constabulary] to act against them."[26] At times, the state itself had even contributed to the journeymen's self-regard. In 1727, the Council of State freed papermakers and their journeymen from the burdens of service in the *milice* (militia), the collection of the *taille*, and the lodging of troops. Furthermore, it limited the latitude of the bosses in their selection of apprentices and granted the company of veterans of each mill two thirds of the 30 livres paid by any "outsider" who wished to

serve an apprenticeship among them.[27] Three years later, in a decree restricted to the généralité of Limousin, Versailles again ruled that the *maîtres-fabricants* could only engage "strangers" as apprentices "in the absence of the sons of journeymen."[28] The state reversed course in 1739, allowing the masters to choose anyone they considered "fit" for indentures. If the veterans protested, they risked a fine of 20 livres per man.[29] Such repeals, however, were unlikely to shake the journeymen's sense that control of the labor market was part of their patrimony—a privilege, after all, that even Versailles had once endorsed. Years later, the Montgolfiers muttered that the journeymen permitted "only the workers of their choice to work in the various mills."[30]

For the most part, the chosen came from paperworker dynasties. Apparently, French paperworkers practiced an almost tribal exclusiveness, leading one pair of scholars to speak of "*l'herméticité* of the papermaking clan" in the Auvergne, where a peasant's son paid three or four times the apprenticeship fee extracted from the offspring of a paperworker. Four generations of ancestors in the craft were necessary for admission into the religious confraternity (*confrérie*) of Ambert's journeymen paperworkers.[31] The men engaged in the trade in Angoumois reserved apprenticeships for their sons and brothers, and "formed a race distinct from the population in the midst of which they lived."[32] Any paperworker born in the nearby Auvergne had to pay off Vidalon's veterans to join their ranks.[33] Moreover, the Montgolfiers complained about hands who refused to labor without compensation beside skilled men who had not been born into the craft.[34] Even the millmasters, said the journeymen, had to possess the proper pedigree. If he was not the son of a master, the man who purchased or supervised a fabrique had to fork over 60 livres to the workers.[35]

With their swagger and cohesiveness, the paperworkers would seem to have been natural claimants to a place in one of the *compagnonnages*, the semiclandestine labor organizations of Old Regime France. Certainly, each of the three principal brotherhoods of *compagnons* welcomed a wide variety of tramping trades.[36] Yet, the journeymen paperworkers, possessors of a distinctive skill, enmeshed in close-knit family webs, and wayfarers on particular tramping trails, remained aloof. They preferred to conduct their affairs through the confrérie, the religious associations that were generally linked to a single trade.[37] Still, paperworkers wandered endlessly. Theirs was a seasonal craft, where a frozen millwheel or dry streambed compelled them to move on. For paperworkers, time on the road was a dusty routine punctuated by welcome inns, chance banquets, and joyous noise—if, that is, the itinerant had been initiated properly. Here was an essential source of the paperworkers' custom, the "conventions made among themselves which they call *modes*."[38]

Through these rituals, feasts, traditions, and proscriptions, the life-cycle of the trade, its ties of acquaintance and experience, and the daily *comédie humaine* of work, meals, personal tragedy and delight were acknowledged and endowed with meaning. These practices reconciled the generations in a shared sense of expectation and privation. The "modes" transformed the commonplace passages of the craft into celebrations, enriching the whole community of workers (sometimes at the expense of an individual) when an itinerant arrived or a coucher traded his easel for the vatman's molds.

The novice learned the "modes" along with his craft. The *droit d'apprentissage* (apprenticeship fee) evidently originated as the journeymen's compensation for the clumsiness of the new boys and the hours spent teaching them the tricks of the trade. The Montgolfiers charged, however, that their old hands taxed newcomers repeatedly, even when the youths had spoiled nothing and taken none of their time.[39] Another manufacturer maintained that the journeymen deserved no reward at all for hours or money lost training greenhorns. Such indemnities, he flashed, were "legitimately due to the master," since "no worker has ever taken the pain, even once, to demonstrate the craft to [an] apprentice." It was certain, he concluded, that if a master and novice compelled a journeyman to do so, the manufacturer's mill would be blacklisted by the workers, he would be fined by their organization, and the apprentice would be "mistreated," that is, battered.[40] When the new man completed his indentures at Vidalon-le-Haut, the journeymen picked his purse once again, this time for the honor of joining them at table (*droit de s'asseoir à table*).[41] While the journeymen savored these gleanings, the veterans also demonstrated to the newcomer that he was under their tutelage as well as the Montgolfiers'. Thus a Parisian apprentice courted trouble when he refused to open the doors for the journeymen paperworkers one morning, "as is customary."[42]

The "modes" dogged the newly minted journeyman's every footstep. When his boss advanced him from the layman's plank to the coucher's felts, his shopmates' approval of the move cost him a sou or two (*droit de changement*).[43] When he landed work, the journeyman had to buy his welcome, in the form of a round or more of drinks known as the *bienvenue*.[44] Here was a material measure of the new man's desire to belong. Treating also reconciled the journeymen's wanderlust and exclusivity: presumably the newcomer was compensating the veterans for the disappearance of an opening that might have been filled by a son or brother. Meanwhile, in a dark corner of a nearby tavern, *les anciens* took the opportunity "to indoctrinate the newly arrived man." Their message was unambiguous: the man who had just bought his welcome

must "swear to be attached securely to their statutes, and in every case to sustain their interests in preference to those of the master."[45]

The fledgling journeyman already knew that his mates took their play seriously. Certainly, paperworkers liked to let off steam: on St. George's Day in 1774, a group of these men from Thiers spent an entertaining afternoon spraying dancers with water from a stream.[46] The "modes" set the rules for this unruliness. They expressed the workers' Rabelaisian fraternity: the Dauphinois journeyman who failed to hold his wine bought the next round.[47] But these customs also reinforced the bonds of honor, family, and identity in a tramping trade, where even short treks brought men into "foreign" parts.

Engagement, marriage, the birth of a child, the death of a companion, or the title of godfather all drained the journeymen's pockets of a few sous, since the proud father or just-married coucher had to fill "many hearty forks and sturdy gullets."[48] Yet, the "modes" also placed limits on this bacchanalian revelry: the Dauphinois journeyman who consumed fat on Friday or missed a holiday or Sunday Mass had to compensate his fellows.[49] (To the Montgolfiers, no doubt, these levies were merely evidence of the journeymen usurping the proper domestic authority of the master craftsman.) The paperworker in Dauphiné who chased after a girl and failed to marry her risked a fine, much as the German journeyman paperworker whose wife bore him a child within nine months had to compensate his comrades.[50] Theirs was a moral community, claimed the French paperworkers, and they were perfectly capable of creating and policing its standards themselves.

They definitely set high standards for their meals and larded them with custom. Since many mills were located at the outskirts of hamlets, it is not surprising that papermakers were expected to nourish their workers. On the New Year, the maître-fabricant in Dauphiné had to provide a *"coq d'Inde"* as well as "the ordinary meal." On Mardi Gras, it was a pig's ear; on Fat Thursday, a ham and glazed crusts; on Palm Sunday, doughnuts; and on Good Friday, a carp. "All of this," an anonymous mémoire spelled out, was an "institution" preserved "in the workers' calendar." Should the millmaster fail to offer these delicacies "on the prescribed days and hours," his entire work force "in concert" demanded their discharge.[51] For Lescourre of Libos, the "best means" for revival of order in the craft was "to oblige" the journeymen to feed themselves: "It is food," he concluded, "which gives birth to almost all the pretexts for the murmurs, cabals, and revolts that one sees holding sway over this species of people."[52] In some cases, however, meals alone drew reveling paperworkers back to the mills. Every time the idea seized them, the angry chronicler sneered, the paperworkers of Dauphiné quit work and sauntered

to the cabarets to drink the day away. They only wandered back to their bosses at mealtimes, "s[a]t down at table as usual, dr[a]nk and [ate], and then they return[ed] to the tavern."[53]

The Montgolfiers summarily dismissed the "modes" as nothing more than excuses for besotted journeymen to reach into their brothers' pockets and waste the proceeds on wine. To another observer, the paperworkers' practices were "irregular, surprising, unbearable."[54] Yet, these customs took shape in an unyielding orbit of hard work and hard play, momentary gluttony and long days of pinched guts, and short lives. A report from 1813 explained that paperworkers lost the capacity to ply their craft at an early age, "their arms losing all suppleness when they attained the age of 42 to 45 years."[55] Men in such circumstances surely exploited their "ways" to the fullest. But these conventions also permitted the journeymen to offset some of the privations of aging, elude some of the consequences of short papermaking seasons, and regulate some of their own passages through the routine trials of their trade. Through their "modes," moreover, they won a remarkable amount of control over the labor market, especially the ebbs and flows of tramping.

For the goldsmith or cooper, the production of a masterpiece demonstrated his skill and, implicitly, his knowledge of the ways of his craft. The strolling paperworker was not afforded this luxury. Immersed in a complex division of labor, he needed to join a vat crew to show his worth. He might possess a certificate of good conduct (a written discharge from a previous master) but that would not impress his new mates. To gain their confidence, he had to demand the customary "allowance" granted to itinerants (*lever la rente* or *droit de passade*). As soon as the informed wayfarer made his claim, work evidently ceased and wine appeared—according to one account, a standard four bottles *"mesure de Paris."* They were finished off in a cabaret, away from the bosses' ears, where the newcomer would "recount in his own way all the vexations that he gave to [his] last master."[56] Worse yet, grumbled one official, the custom rewarded the most ungovernable man, who was moved "by mutiny or by profligacy" to abandon his employer or who had to be sacked. Then this undisciplined *garçon* was free to wander the "length of the river" and disrupt work in mill after mill until he was sated.[57]

There was, of course, more to the droit de passade than drink. The man who called for his rente undercut the wariness prompted by a distant accent. Essentially, the custom was a means of "legitimate" foraging for men ceaselessly on the tramp. Each of the journeymen the itinerant encountered, said the prefect of Puy-de-Dôme as late as 1825, was expected to offer him a subsidy of 10 *centimes*; each master was supposed to provide five.[58] This advance refreshed a tired traveler, reinvigorated him before the boss tested his skill, or

replenished his purse before he decamped in search of a soft foreman. Meanwhile, he enjoyed about a week of lodging and work. If there was an opening, the millmaster might even suggest an extended stay and a particular wage. To one small producer, the purpose of this "right" was devastatingly clear. The wayfarer avoided the humiliation of requesting work; instead, the boss humbled himself by making "proposals" to the newcomer.[59] By turning the tables, the itinerant got the manufacturer to validate his skills, whether or not he possessed the references mandated by the state. Moreover, the veterans already in the papermaker's employ preserved a say in hiring, since they likely taunted or beat into a hasty retreat the man who failed to claim his rente. Finally, the easy access to the shops enjoyed by itinerant paperworkers revealed how tenuous the masters' control over tools and men really was. In 1785, a petty Auvergnat manufacturer had no choice other than to permit *les rentes* (as the men who traded on this custom were termed) to "pass and pass and even pass again" through his shops.[60] He lacked the strength to stand alone against the combined journeymen paperworkers of his region, and he knew it.

The papermaker who resisted the "modes" soon found his mill "damned."[61] His own hands quit en masse, and only the desperate or foolhardy journeyman dared to replace them. The ban might last as long as six months, and when it expired the workers had "the cheek," at least in Dauphiné, to demand compensation of 30, 40, 60, or even 100 livres from the willful master. (He had, after all, violated their "statutes" and prevented them from earning a living.) Should the boss refuse to pay immediately, the journeymen threatened to shutter his mill "forever." Even the most resolute manufacturer, lamented the Dauphinois chronicler, inevitably capitulated and bought off the veterans.[62] Assessing the frequency of such episodes is not possible, but one, admittedly irate, producer believed that the journeymen's "fines and other arbitrary taxes" reduced France's annual output of paper by 12 percent.[63] "Here is a very strange police," the Dauphinois mémoire deadpanned, "in a kingdom which prides itself on maintaining proper order [*bon ordre*]."[64]

As the journeymen paperworkers patrolled the borders of their republic, they "damned" their wayward comrades as easily as the bosses. For men on the move, mastery of the labor market was crucial. Paperworkers enjoyed this dominion and trusted in sanctions and blacklists of their own making to sustain it. They forced the "foreign" journeyman to contribute handsomely to his new confraternity's coffers. They "expressly prohibited" any in their company from toiling beside a worker or apprentice who had been "proscribed, banned, or fine[d]" until the miscreant had paid the price—literally. They taxed any journeyman who requested work "secretly." And they punished the

worker who made a personal profit from a penalty: these assessments were to be relished in buckets of drink and swollen guts "in common."[65]

Versailles attempted to lay down the law to the ungovernable paperworkers in a series of edicts that stretched back to the late seventeenth century. But state penalties against journeymen who fined their masters or passed information about blacklegs and strong-willed bosses to nearby mills were little more than paper tigers. "For the paperworkers," an Auvergnat official admitted, "the [government's] regulations are nonexistent."[66] Small wonder that the Dauphinois reporter spat: "What excess! What conduct! What patience on the part of the informed administration!"[67] Typically, Mignot, an Auvergnat *subdélégué*, thundered in 1772 that the time had come to make some examples, but he never backed his verbal severity with action.[68] The Montgolfiers believed that the edict of 1777 outlawed apprenticeship fees, the workers' "police" of their brothers, the introduction of journeymen into the shops without the master's permission, the abandonment of work in pursuit of wine, and nights in the cabarets. When these practices cost enough paperworkers "their money and their liberty," they would cease.[69] So the Montgolfiers waited in vain for stern enforcement of the edict.

Such pleas may have fallen on deaf ears because most manufacturers did not want their own skilled men jailed, and the government did not wish isolated incidents to become river-wide rebellions. Consider the "disorders" in Castres and Burlats, two Languedocian towns, in 1786. On May 9, the journeymen, aware that the pulp was ready to be worked, "suddenly" threw down their tools. They demanded a wage hike and decreed the substantial penalty of 60 livres against any man who continued to labor in the fabriques. They wrote to the workers of the other mills in the province and beyond to avoid their town until the masters caved in. On June 7, the *maîtres papetiers* of Castres requested a pardon for their delinquent hands; doubtless, the two sides had reached an accord, and the state had no interest in spoiled pulp.[70]

"The insubordination of the workers," wrote an inspector of manufactures in Franche-Comté, "is general in the profession [of papermaking]."[71] In Champagne, a shaken intendant argued in 1783 that "the spirit of association and of mutiny which had always been peculiar to the journeymen paperworkers is thrust to such excess today that the severity of the ordinances seems insufficient to stop the disorders."[72] Ever fearful of renewed rebellion, the intendant of Languedoc concluded in 1786 that the "paths of gentleness [*voies de douceur*]" were the best to follow in dealing with the indomitable paperworkers.[73] Nor is it surprising that the authorities believed that the thirty-five thousand or so French paperworkers of the late Old Regime were closely united in regional and even national combinations.[74] In 1777, the Council of

State maintained that "the workers of the kingdom's paper mills are bound by a general association." So powerful was this body, the government continued, that it rendered the journeymen the "masters of the success or of the ruin of the entrepreneurs." In fact, the workers' "so-called jurisdiction" portended the "total subversion" of the craft.[75]

Still, it is worth noting that the journeymen paperworkers of Ambert had "outlawed in perpetuity" those of nearby Thiers; "in the spirit of reprisal," the workers of Thiers returned the favor.[76] The semiclandestine nature of the paperworkers' combinations limits our knowledge of their actual range. Surely, Etienne Montgolfier had reason to claim that these men "always" focused "alert eyes [*yeux ouverts*] on every paper mill of their province, and even of neighboring provinces." But his contention that "the paperworkers of the provinces of Languedoc, Auvergne, Dauphiné, Franche-Comté, and Bugey form an association among themselves" probably stretched the workers' effective reach.[77] More common, I suspect, was the "cabal" organized by the hands in several Dauphinois mills to augment their wages.[78] As one informed observer explained, the journeyman who violated his brothers' rules could compensate them, abandon the trade, or "leave the *pays* [region]."[79] Consider, too, that a vexed subdélégué portrayed the temporarily united paperworkers of Thiers and adjacent Clermont-Ferrand as an "unworthy [*indigne*] nation."[80]

A recent history of modern French papermaking echoed earlier claims that the journeymen were increasingly insubordinate during the last years of the *ancien régime*. The general revival of the industry, the establishment of large, new mills around Paris, and the journeymen's successful pruning of their ranks offered the workers greater leverage—which they did not hesitate to use.[81] Perhaps incidents were more frequent on the eve of the Revolution, but if so, the journeymen were drawing on a long heritage of strikes, work stoppages, and noisy displays. In 1600, for example, the papermakers of Vieux-Thann and Cernay, in Alsace, fumed that their workers were "always insatiable; they are never satisfied with the food that is served to them; if one does not give in to their caprices and to their effronteries, they quit after a warning of eight days, leaving in the lurch, without hope, the master and his mill."[82] In 1732, the masters of Ambert complained about a rising that turned on the journeymen's right to wages when insufficient water closed the mills.[83] And a strike in 1767 stemmed from the papermakers' decision to provide a stipend of 30 sous per month to the workers in place of their regular ration of wine, which had grown costly.[84] If a pattern emerged at all among these strikes and *émeutes*, it was the recurrent efforts by calculating producers to hold down costs by casting out a custom or two. Since the journeymen defended

their ways ferociously, such attempts inevitably provoked bitter disputes. Whether it was a matter of wages or techniques, unsigned letters tucked in a child's pouch carried the news from vat to vat. As Etienne Montgolfier explained, the journeymen did not suffer "the slightest change" unless the producer had the approval of his own workers as well as that of their entire association.[85]

The proud, self-reliant paperworkers, *"ce corps républicain,"* raged one official, remained "jealous of a self-styled, chimerical independence." Even worse, he concluded, was the journeymen's willingness to seize "the occasions to signal" their clout.[86] After an incident, the Dauphinois workers expected their master to "take the first steps; in effect, it [was] always he who [went] to search for them at the cabaret."[87] The Thiernois strike of 1772 ended only when the manufacturers summoned their hands and paid their wages for the two months of the walkout.[88] No doubt, such concessions were especially loathsome to master papermakers who characterized their workers as "rabble," a "wretched race," and most vividly, "urchins that the Auvergne vomited out to torment the Dauphiné."[89] As one Auvergnat papermaker maintained, "It is only the indigent and unfortunate class which takes up this craft."[90] Yet, it was these base creatures who exercised, through their skills and the custom that shielded them, their topsy-turvy dominion over the craft. "Impunity," Etienne Montgolfier stormed, "has emboldened them." The journeymen's "principal goal is to suffer no change nor ameliorations in the mills where they work and to maintain in them the ancient customs or *modes*," this level-headed man fumed.[91] He realized that his family would have to expel the journeymen's whole custom from their mill if they intended to match Dutch techniques and govern skill—and the skilled—themselves.

CHAPTER SIX

The Lockout

Donning green aprons, paperworkers loyal to the glazing hammer once battled across Europe with the hand-burnishers, who signaled their defense of time-honored practice with gray or brown aprons.[1] Such resistance to technological innovation was commonplace in hand papermaking. Pierre Montgolfier's journeymen, too, had contested the introduction of the Hollander beaters at Vidalon-le-Haut in the 1750s. By the early winter of 1781, however, the thump of his mill's beaters revealed that the device had finally secured a firm footing on French soil. Even more threatening to Vidalon's veteran hands, the Montgolfiers were engaged in "the training of workers, fifty new paperworkers."[2] Such a resolute challenge to the journeymen's custom, however, had its risks. A Breton manufacturer was injured in 1786 by journeymen angered by his willingness to hire a "capable" paperworker whose father toiled as a hatter. Although the skilled man's mother sweated in a paper mill, he still lacked the appropriate lineage—and perhaps familiarity with the "modes."[3] So the Montgolfiers were courting danger when they selected for their expanded shops young men free of ties to the paperworkers' dynasties—and ways.

The Montgolfiers intended to unravel the fabric of acquaintance, experience, kinship, skill, and work itself that the "modes" had woven together. From one angle, these customs amounted to a sort of subcontracting, through which the journeymen set standards for hiring and firing. The Montgolfiers attempted to overthrow this regime. They alone would incorporate workers into their internal labor market, determine the newcomers' skills, and assign them jobs. No longer would the trade be passed "de mâle à mâle," that is, from one generation to the next, with the journeymen's consent.[4] Thus the Montgolfiers never considered the wholesale retraining of their veteran hands. They were transforming new men into a new kind of worker.

German paperworkers, it was said, permitted "nothing new to be introduced nor anything traditional to be discontinued."[5] Like their German counterparts, French journeymen resisted novel technology, and Dutch practice did constitute genuine, if limited, reform. But the basic division of labor of the

trade and even its nomenclature remained intact. Accordingly, continuity and control, rather than the intricacies of the Dutch procedures, were the central issues at Vidalon-le-Haut in 1781. For the journeymen, familiar skills were the touchstones of a system of meaning, legitimation, and raucous noise, and of bonds between fathers and sons. Moreover, they were enduring measures of a mastery codified and celebrated in the journeymen's own "statutes." Consequently, even the technical amendments of the Dutch approach disturbed Vidalon's journeymen. They did not, however, engage in Luddism. Instead, the revolt of Vidalon's veterans in 1781 was sparked by a deeper concern and a frightening alliance: the Montgolfiers' willingness to stand by an apprentice who refused to pay his droit d'apprentissage.

At a Flemish mill, Ecrevisse, Desmarest's man in Annonay, had learned that one "did not have to draw workers from Holland to make papers like them."[6] He had built a work force from scratch, with native sons as his raw material, and they turned out first-rate wares. In his early days as a journeyman paperworker, he had been a model of "rectitude."[7] As his reputation as a millwright grew, Ecrevisse dreamed of renting a "small *fabrique*" where he could organize production (*faire des opérations*) as he saw fit, work at the "perfection" of his shops, and dedicate himself to the study of his art "without constraint."[8] Instead, he got the chance to exercise his vision of fabrication as applied science at a very large mill. Even more, he had a large part in training the Montgolfiers' *"bons élèves,"* the fresh hands free of the workers' custom that he had apparently avoided during his years as a journeyman.[9]

Desmarest's advice to surround the beaters with new hands and Ecrevisse's success with novices doubtless heartened the Montgolfiers. "The best means" to put an end to the journeymen's "modes," Pierre Montgolfier believed, was "to render the workers more dependent."[10] This dependence, he assumed, would break the ties among workers and encourage bonds between master and men. In 1769, he had barred tramping men from "raising the rent" at Vidalon-le-Haut.[11] Word of his boldness had flashed across the journeymen's grapevine, and the former Montgolfier hand who found work in Dauphiné had to pay for Pierre's steeliness: the fine was 36 livres.[12] In fact, Pierre's own journeymen may have been somewhat less headstrong than their brothers elsewhere. Certainly, the paperworkers of Thiers considered their counterparts in the Vivarais to be "too docile towards their masters" and balked at toiling in their company.[13] (Of course, the Montgolfiers' journeymen returned the favor and extracted compensation from the Thiernois worker who joined them.)

But the Montgolfiers' veteran hands still had backs too stiff for their masters. Pierre launched frequent tirades about journeymen "working only by

routine, [who] have the vanity to believe [they] know all and will not learn."[14] Etienne spoke of paperworkers carried away by "excesses that one would hardly believe."[15] Vidalon's *patrons* stormed about journeymen who "refuse to work at the task suitable to the master and want to select their spots"; who "immerse themselves in the disputes between a worker and his master"; and who "all threaten to abandon their work every time the master wants to establish order [*mettre l'ordre*] in the mill."[16] The Montgolfiers had had enough. Yet, peopling their mill with greenhorns innocent of the "modes" was no easy task, as the Montgolfiers were well aware. The master papermakers who had dared to step beyond the bounds of the journeymen's combination and mold "docile" workers "saw those of the Association who are employed in other mills gather, menace, injure, and attack the pupils [*élèves*] whom they were training."[17]

Meanwhile, Vidalon's *patrons* had learned to count on little from their fellow producers: "If all the manufacturers behaved well and carried out the Regulations, that would be sufficient; but many among them, either due to weakness, ignorance, or even jealousy, tolerate or even foment this spirit of insubordination." Rather than risk "the momentary loss" of materials and men, these producers "submitted to the [workers'] yoke and are bound by their ideas."[18]

The Montgolfiers were confident in their ability to transform Vidalon-le-Haut into a beneficent institution, a "temple of rational technology" rich in skilled, malleable workers.[19] Certainly, their efforts mirrored the Enlightenment's faith in reasoned design, in the capacity of a carefully tooled environment to turn out better wares and workers. As their collaborator Desmarest explained, "The first thing in an enterprise . . . is to know where one is going and if one can arrive there."[20] Yet, the Montgolfiers' recourse to fresh faces, to a work force of tabulae rasae in order to solve their labor problems, was also a function of memory and experience. It was the complex texture of power in the shops and conflicts as old as the craft that pressed the Montgolfiers to "form" new men. And it was the desire to deregulate their elevated sphere of the industry that fed their plans. Their nouvel ordre was not simply an Enlightened abstraction at work; it was also the residue of a rough-shouldered struggle between battle-scarred journeymen and entrepreneurs.

These scars originated in the lesser Montgolfier mills as well as at Vidalon-le-Haut. In 1776, the journeymen blacklisted the small Montgolfier mill in Rives, perhaps because the provincial government of Dauphiné had encouraged technological reform there. One worker, Bession, remained loyal to his employers but at great expense. His treasonous fidelity was met by a fine of 30 livres and, evidently, the threat of rough handling by his fellow journeymen.

He decamped to Thiers with his certificate of previous employment in good order and soon found work, but his new mates abandoned their tools once they learned Bession's story.[21]

Augustin Montgolfier, who owned the mill at Rives, was habitually high-handed with his workers. Yet, this short-tempered man also shared something of his brother Etienne's calculating nature, as an incident in his shops in February 1781 and its aftermath revealed. While Augustin was in Geneva on business, his wife's Sunday meal with "the most distinguished people in the locality" was disrupted by two angry workers. They had arrived after their comrades' supper, and though a serving was offered to them, their fury spilled over in thrown plates, insults, and expletives. Joseph Montgolfier, Augustin's brother and the proprietor of a paper mill in nearby Voiron, restored peace by chasing the two journeymen out of the mill. On the following day, the two workers returned, demanded that their accounts be drawn up, and began a tour of the neighboring mills, airing their grievances. The paperworkers' association struck quickly, condemning the two workers for not seizing their betters' food for themselves. It also levied fines against Augustin's wife, Joseph's millmaster, and the journeymen of both mills for their failure to stand with the dissident pair and their willingness to return to their benches after the two men had taken to the road. Furthermore, Augustin's and Joseph's hands had to abandon their toil until they paid off their brothers, or risk the blacklist and permanent exile from the métier. A distraught Augustin found "all [his] works idled" when he arrived home, and "despite the ridiculousness of the demand, [his] workers as well as those of [his] brother all preferred to leave without their arrears rather . . . than get themselves in trouble." Instead, they left Augustin in trouble, in the form of three thousand pounds of perishing rags.[22]

Nevertheless, the veterans' flight provided Augustin and Joseph with just the opportunity they were seeking. They decided to replace their journeymen with carefully selected and trained youngsters uninitiated in the veterans' ways. Six men of the paperworkers' association gathered swiftly and marched on Voiron, said Etienne Montgolfier, bent on "insulting and mistreating the new workers."[23] Three were arrested and jailed in Grenoble, where they remained in December. Meanwhile, the paperworkers' combination took care of its own: it imposed a levy of 15 sous per month on each member to sustain their imprisoned mates.[24]

In September, Augustin explained the course he and Joseph had taken in letters to his family at Vidalon-le-Haut.[25] Assertive and self-congratulatory, these dispatches served, in part, as blueprints for the lockout at the big mill in November. Be decisive, sure, and unshakable, Augustin instructed his father and brothers. He believed that his nouvel ordre, shorn of the "modes," would

even find favor among many experienced hands: "Half of the workers themselves may want this order and . . . they only refuse to submit to it out of fear that we cannot always maintain it."[26] Convince your workers that you are determined to establish a new regime, Augustin exclaimed, and you will see many renounce the "modes" and embrace your new way.

Augustin was equally sure about the best approaches for the "formation" of the fifty fresh hands who thrilled Etienne. Set up vats manned solely by these tyros, he counseled. Could two distinct labor forces exist within one mill? Make sure, Augustin continued, that the new men and veterans ate and slept apart, and "watch with the greatest attention in order to avoid all communication" between them. As the youngsters gradually mastered the art, they would bridle the pretensions of the veterans. He seems to have assumed that the indomitable journeymen would tire of all this and depart, while "the good, old subjects" would be encouraged to stay. As a last swipe at the ungovernable men, he even suggested that Vidalon's *patrons* staff a vat or two with women. They "cost much less and, with assiduous work, could produce as much as the men."[27]

"Every imaginable vexation," Augustin warned, would accompany his family's imposition of a new regime. Former hands would threaten to "set fire to the four corners of the [mill]," the journeymen's wives would attempt to lead the new men astray, and their children would soil the rags and vats. Despite these burdens, Augustin crowed, he had prevailed. And he depicted victory in terms bound to delight Vidalon's bosses: he had produced "much tranquility" in his shops; he had "closed the door to all the *Rentes*. . . . There are neither *rentes*, nor fines, nor *bienvenues*, nor *associations* of any kind."[28] But when he moved his shops and imperious ways with workers to the Beaujolais in 1785, in order to produce his wares in a province free of the paperworkers' association and to sell his reams throughout the Lyonnais without duties, his scrupulously reconstructed and enlarged mill burned to the ground.[29]

Pierre and Etienne Montgolfier were having their own troubles with Vidalon's journeymen. "For two months," one of Vidalon's *patrons* reported in May 1781, "I have been struggling to establish an order [at the mill]. I lost several workers, but happily they are neither the most well-behaved nor the best."[30] In October, father and son evidently chose to follow Augustin's counsel and take the offensive against the journeymen. They "wanted to reform several abuses which were hindering the introduction of the Dutch procedures and machines." In response, Vidalon's veteran hands assembled several times, laid down their tools, consulted with journeymen in neighboring mills, and finally returned to work.[31] On October 29, the workers taxed the appren-

tice Poynard, one of Etienne's prized newcomers, who refused to pay.[32] (The Montgolfiers, for their part, claimed that Poynard had simply "renounced one of [the journeymen's] preposterous laws.")[33] Vidalon's veteran workers then walked out together and sought refuge in a cabaret in town. A day later, when the greenhorn kept his purse closed, the journeymen refused to pick up their tools and set the discharge of the errant youth as the price for their return to work. When Pierre Montgolfier rejected their bid, the journeymen stayed away.[34]

The Montgolfiers soon realized that their veteran hands would not come back to their molds and felts when the "fumes of wine" had passed.[35] On October 31, they asked Mathieu Johannot, in his capacity as local warden (*maître-garde*) to write up a complaint so they could place their case before the intendant. Johannot advised the workers to resume their toil, which they did, but only after they had declared that they would abandon Vidalon-le-Haut en masse in six weeks, the proper term of notice mandated by royal decree.[36]

On November 2, two journeymen, Androl and Palhion, demanded that Jean Châtagnier, another of the Montgolfiers' fledglings, pay his droit d'apprentissage. He demurred, with Pierre Montgolfier's backing. The journeymen again tossed away their tools, and they also apparently menaced and insulted Jean-Pierre Montgolfier. The workers returned to their benches on the following day, but Androl and Palhion, who had been instrumental in the effort to exile Poynard, were not mollified.[37] On November 7, they quit their work, mocked Jean-Pierre Montgolfier, and set off a din at the dinner table. Later that day, Androl, who had evidently returned to his vat, let the support of a press slip. The pressbar bowled over an apprentice, wounding him in the groin. Androl and Palhion then hectored Jean-Pierre Montgolfier to replace the injured boy, cautioning their boss that they would fall on him and beat his "ass" (*bosse*) at the slightest provocation. That evening, the worker Marcheval, armed with a rock in his hand, taunted Pierre Montgolfier, threatening to crush the old man like a "toad" if he raised his cane against him.[38] On Pierre's complaint before the seneschal's court in Annonay, on which yet another of his sons, Alexandre-Charles, sat, two of his "most mutinous" hands were arrested and briefly imprisoned. In December, true to their word, Vidalon's veterans quit the mill, but not before barring it to fellow members of their association.[39] Several lingered in town and, on a March night in 1782, apparently battered three of the Montgolfiers' new men.[40]

On January 22, 1782, the Montgolfiers declared victory. Aided by five veteran hands who had straggled back to their mill as well as approximately twenty novices, they had six vats turning out paper. "I am content enough

[with this production]," wrote one of Vidalon's *patrons*, "to congratulate myself every day for the course I took; several of my trainees are working as well as [had] the great majority of the *ouvriers de la mode*."[41] The incidents at Vidalon-le-Haut in the late autumn of 1781 constituted a lockout precisely because the Montgolfiers knew that their veterans' only choice was to follow their "modes" into exile. The mere presence of the apprentices under Ecrevisse's wing, recruited among the sons of day laborers and vine-dressers, must have infuriated Vidalon's journeymen. Untouched by the "modes," these youths amounted to a daily demonstration of the veterans' loss of mastery over hiring, the mainspring of their powers. Even after they gave their collective six weeks' notice, the Montgolfiers' journeymen continued to press for apprenticeship fees from the novices.

Certainly, Etienne Montgolfier recognized the challenge posed by his new hands. "In the whole province of Dauphiné," he explained, "no apprentices [were] trained, which produce[d] a dearth of workers." The result was plain to see: Dauphiné, "recruiting [journeymen] at the expense of its neighbors, [was] also the province where the workers exercise[d] their police most freely and indulge[d] in the greatest excesses." If skilled men became less scarce, the "modes" would become more so. "Determined to maintain order and subordination" in their mill, his family trained "newcomers [*nouveaux élèves*] to replace the old workers."[42] They had "smashed the yoke and formed new workers." Here was the solution to the problem of overmighty skilled men: "I presume," one of the Montgolfiers reasoned, "that the other manufacturers will have to follow the same course, if the veteran workers do not change their ways."[43] Accordingly, Vidalon's *patrons* proposed that each papermaker train a cohort of apprentices in proportion to the scale of his works. Only then could their confrères share his joyful report that Vidalon-le-Haut was "worked almost entirely by new workers who have nothing in common with those of the Association."[44]

In the weeks after the veterans vanished, the Montgolfiers themselves toiled around the vats.[45] More valuable, surely, was the willingness of certain Ambertois producers to help them fill their orders.[46] This solidarity among manufacturers (which was not always the case) and the presence of their new hands doubtless braced the Montgolfiers' resolve to extinguish the "modes" in their mill. So, too, did their patrimony of innovation, that is, their vision of an industry shaped by applied science and the shifting taste of consumers. Still, the journeymen's culture they contested was not a "pre-industrial," pre-market inheritance. Rather, the "modes" had evolved in a capitalist system of production; indeed, they permitted the journeymen to take advantage of and even shape the labor market, while cushioning its vicissitudes. Thus the workers

sought to preserve the technology that underscored this mastery. And the Montgolfiers, eager to perfect their art, had to oust the "modes" to take command of both skill and the shopfloor. Here was the impulse of their nouvel ordre, a regime determined less by the arrival at a particular stage of industrial development or the imperatives of new machines, and more by the enduring realities of skilled production and the embers of the protracted struggle that culminated in the lockout of 1781.

PART THREE

Managing to Rule

CHAPTER SEVEN

The New Regime

"We have stepped forward beyond the other manufactures and we must be content to train up hands to suit our purpose."[1] These were the words of Josiah Wedgwood, but they could have been spoken by Matthew Boulton, Christophe-Philippe Oberkampf, or the Montgolfiers. Vidalon's *patrons* had stepped forward by introducing Hollander beaters and fresh workers into their shops. Both new machines and new men fed the Montgolfiers' sense of mastery, yet they remained wary of the men of the "modes" and dependent on Ecrevisse's "enlightened pains."[2] As producers of elegant, expensive papers they faced particular problems. Consider the complaint, for example, of an indignant subscriber to the quarto *Encyclopédie*: "The paper is generally defective and the type almost faded, which greatly fatigues the reader's eyes. . . . The major portion of the sheets [of paper] are blurred or torn."[3] To satisfy their discerning, elevated clientele, the Montgolfiers had to turn out large batches of high-quality wares and do so with young, inexperienced hands. Yet, they were more than content to "train up" their company of novices, a cohort, they imagined, that would permit them to mold their mill into a theater of innovation and profits.

Grooming a work force of newcomers was expensive business, Etienne Montgolfier understood, but he had not "feared" this cost. Instead, he "sacrificed his momentary interest" to perfect his technology and trainees. The whole craft would gain from his family's approach to labor discipline, just as it profited from their installation of effective beaters. His "new shop had always been open to the manufacturers of the province who desired to obtain enlightenment there." Furthermore, Etienne himself had sketched "exact drawings" of his "machines and of their diverse details," and he presented them to Desmarest, who intended to make these diagrams public.[4] Vidalon's *patrons* took the same tack with their disciplinary regime, believing that it was their responsibility to school fellow papermakers, hostile paperworkers, and their own raw recruits in proper conduct and practices. When Jeanne Brialon fled Vidalon-le-Haut in 1785 without her discharge papers and without offering proper notice, Etienne transformed the incident into pedagogy.

If Brialon's new boss, a widow named Palhion, failed to return her, Etienne threatened to seek the stiff penalties prescribed in royal decrees—a fine of 100 livres against Brialon and a levy of 300 against her mistress. He justified his resolve by pointing to "the sacrifices [he] made to establish order and the subordination of the workers in [his] mill." Every papermaker would benefit from his daring; none would be allowed to obstruct his success: "Madame Palhion must appreciate that [he] will not allow the [workers'] abuses to be reintroduced [at Vidalon-le-Haut]." He vowed to pursue Brialon, if she took flight again, until he lost her trail. He would then obtain a judgment preventing her from toiling in the "*pays* or the environs."[5]

Brialon soon returned to Vidalon-le-Haut. The Montgolfiers waived her obligation to remain for six weeks, granted her formal discharge papers, and sent her back to the widow Palhion. At the outset of this affair, the Montgolfiers declared, "This case will serve as an example for the others."[6] At its close, they contended that they had hounded Brialon "to prove to the workers that they will be punished for improper conduct."[7] Heedless manufacturers, the Montgolfiers' stern posture suggested, would suffer the same fate.

Yet, the Montgolfiers' unflinching pursuit of Brialon can also be read as the handiwork of masters almost too eager to display their newfound mastery. The program of labor discipline they designed and imposed at Vidalon-le-Haut was anything but prefabricated, ready for use just as soon as the mill's veterans moved on. It was an experiment or, more properly, a set of experiments. Busy-beaver entrepreneurs in the classic mold, the Montgolfiers fashioned one plan of action after another. Some of these protocols covered one or two sheets, usually of paper too rough or blemished to sell; others ranged over fifteen pages. Some were probably posted as mill codes; others appear to be drafts circulated for comment within the Montgolfier family. Most were the products of a single hand, but a few contain amendments proffered by a second writer. The most important measures were the "rules to observe in the paper mill of Montgolfier of Annonay," the "rules to follow in Pierre Montgolfier's mill," the "general rules of the Montgolfier paper mill," "my reflections on the workers necessary for a paper mill," "my ideas on the government of a paper mill where one feeds the workers," the confident "plan to establish order in a paper mill," and of course, the "fixed and agreed upon regulations in the royal manufactory of paper of Annonay."[8]

Formal and precise, these handwritten directives nevertheless reveal a provisional quality. Read together, they possess the feel of restless experiment: should Vidalon's workers arrive for Sunday and holiday suppers during the winter no later than 6:00, as the "general rules of the Montgolfier paper mill" urged, or 6:30, as a second mill ordinance decreed, or 7:00, as a third code

demanded?[9] Whether dissent or mere inconsistency, such variation reflected the evolving character of the Montgolfiers' regime, one drawn from Pierre's catechism classes for young hands, Jean-Pierre's recollections of his tour de France, and Etienne's calculating, mathematical temperament. It was the work of men of principles, but these principles had emerged from hard-won experience—the difficulties of managing a complex production process and conflicts with generations of journeymen—as well as an Enlightened passion for improvement.

The Montgolfiers' unsettled nouvel ordre left little to chance. From hiring to firing, they established rules for every sphere of work and life at Vidalon-le-Haut. Elaborate hiring arrangements introduced the sons and daughters of *journaliers* and *vignerons* to the expectations of their new masters; the mill village and the masters' paternalism offered fresh reciprocities and insulation from the "modes"; the wage system cultivated stability and self-interest; and impersonal discipline spelled out standards and responsibilities. Taken together, these venerable and inventive practices were geared to produce an unconventional sort of paperworker: an employee.

When a compositor in eighteenth-century Neuchâtel or a mason in Renaissance Florence signed on with a master, he had "no notion of joining a firm."[10] He finished a book or a cornice and then moved on, or was sacked. His security in employment rested on his reputation, family ties, and friendships, rather than long-term commitments by the boss. Paperworkers, too, circulated from job to job, lasting only as long as a stream flowed and rags were available. Their masters, moreover, had no interest in maintaining permanent work forces; indeed, most papermakers learned how quickly the journeymen withdrew their labor "for wages or customs."[11] So the Montgolfiers' decision to create a stable company of workers cut against the assumptions and practices of early modern European employers in general and papermakers in particular. Even more, their effort to recruit and train a cadre of men and women with a tractable personality doubtless surprised those familiar with the stubborn, contentious paperworkers. They dreamed of a sort of bound journeyman, the durable possessor of a veteran's skills who would serve as the instrument of rather than impediment to their ceaseless retooling of Vidalon's shops and routines. Thus the Montgolfiers were not engaged in deskilling, at least as it is conventionally understood. Instead, they intended to demystify skill and reduce it to mere manipulation, whose mastery earned a reward from the boss in place of a toast—or a levy—by old hands. Rather than separate conception and execution, they linked them in a new key, and they alone would call the tune.[12]

The novelty of the Montgolfiers' nouvel ordre, then, was its goal. Indeed,

its distinctive feature was the latitude Vidalon's *patrons* enjoyed to impose their will and plans. Put another way, they were in an unusually favorable position to control the production of skill. Nevertheless, even the most advanced industrial management in the era of Enlightenment (an age associated with breaches rather than bridges) preserved as well as swept away. The Montgolfiers' new regime retained the traditional division of labor and its nomenclature, the convention of nourishing workers, and a wage system that mixed time- and piecework. Thus Sidney Pollard misses the point in claiming that "long-term 'bindings,'" such as those that ensnared Vidalon's new hands, harked back to earlier times, while the "'free' labour contract" was forward-looking.[13] The key question was the fit of a particular policy or instrument in the entrepreneur's larger design. Creating a stable company of hands in a skilled craft was surely anything but backward-looking.

The first principle of the Montgolfiers' nouvel ordre surfaced in the first words heard by every new arrival at their mill: workers will not meddle in the affairs of their fellows or their bosses.[14] Vidalon's *patrons* fostered the worker's individual and family interests, as well as his interests as a member of a vat crew, at the expense of wider solidarities. For their part, Vidalon's new model workers applied and resisted the Montgolfiers' disciplinary imperatives. Of course, Etienne's bons élèves did more than respond to external compulsions. Indeed, Vidalon's new hands adapted the nouvel ordre to serve their own purposes.

CHAPTER EIGHT

Hiring and Firing

In June 1785, the Montgolfiers described their labor force to a fellow manufacturer. Their shops were worked by young men they had trained and several old hands who, "tiring of the debauchery of their comrades" but unable to find respite "under the tyranny of the *mode*," had "become attached" to Vidalon-le-Haut. The Montgolfiers reported happily, moreover, that the rentes, the men sustained by this customary largesse, "hardly come by the mill."[1] Word had spread about the Montgolfiers' new men, and their brother producers naturally inquired about securing one or two. (Evidently, these fabricants believed that the hands themselves mattered, rather than the Montgolfiers' regime.) In any event, Vidalon's *patrons* clung tightly to their carefully groomed fledglings, wishing to expose them neither to suspect journeymen nor their ways: "It is not possible," said the Montgolfiers, "to lend you [our] workers to subject them to harassment by the other workers."[2] If their confrères wanted the right sort of hands, they would have to emulate the Montgolfiers' strategy of selecting the proper raw materials and carefully molding them.

As practiced by Vidalon's *patrons*, "the adaptation of the labour force" began before the labor did.[3] The Montgolfiers treated hiring as the first step in their campaign to produce a disciplined company of workers. Theirs was an early attempt to build an internal labor market, a system in which, ironically, market considerations are less important in the price, training, and allocation of labor than administrative purposes and procedures.[4] If successful, the Montgolfiers could avoid competition on the open labor market and the concessions granted to journeymen in periods of feverish demand. Simply put, an internal labor market promised greater prospects for the control of skill—and skilled men—than had fishing journeymen from the pool of itinerants.

In some ways, the world of Old Regime papermaking was a small place. The Montgolfiers could promise a customer in 1780 that "in the future" they would take pains "to entrust the fabrication of [her] tapestry-paper only to those of [our] workers who have the surest hand[s] [*la main la plus égale*]."[5] In the same year, Vidalon's *patrons* informed Augustin Montgolfier that they had learned "from a worker" that a second man had abandoned a mill in

Chabeuil. "We thought," they concluded, "that maybe this man would be suitable for your mill in Leysse [Savoy]."[6] So the Montgolfiers always had leads about which tramping men and which of their own men were most worthy.

It remains unclear how many of the Montgolfiers' hiring procedures took shape after the lockout, but they had certainly acquired a new edge. Forging an internal labor market from untested day workers and vineyard hands put a premium on information. Before they engaged a novice, the Montgolfiers, especially Jean-Pierre, scrutinized his bearing and background; before they added a veteran, they inspected his references. If the greenhorn or grizzled hand passed their tests, the Montgolfiers demanded that he affirm a set of mill ordinances; when the *patrons* remained skeptical, and they often did, they placed the new arrival on probation. Fresh from their struggle to employ both the techniques and the men they desired, the Montgolfiers never tired of demonstrating their authority, particularly to newcomers, who had to distance themselves from the "modes." For example, Jacob Mosnier had to disavow these customs to win a job in 1783, and two years later, Claude Duranton's introduction to the mill included the warning that his new bosses did not tolerate the "modes."[7]

As master craftsmen, the Montgolfiers longed for the proper filial subordination of journeymen and apprentices alike. As Enlightened improvers, they yearned for men who would function as reliable instruments within their calculated designs, without obstructing their technological ventures. As longtime manufacturers, they were well aware of the state's failed effort to restrain the ungovernable paperworkers. "You will find me eager," one of the Montgolfiers exclaimed, "to maintain proper order among the workers and concord among the manufacturers."[8] First, however, Vidalon's *patrons* had to tend to their own garden and, most especially, keep it free of weeds. The crowded Vivarais, one of France's "reservoirs of men," was an ally.[9]

Obtaining an adequate supply of youths for apprenticeships never troubled the Montgolfiers. Throughout the 1780s they promised candidates the second or third spots that opened and turned away others with a pledge to call them when their services were needed.[10] They came, furthermore, despite the threat of the journeymen and their association. As one of Vidalon's *patrons* boasted, "I had, in the first days, a bit of difficulty in procuring workers; but when people knew my project better and when they were convinced that I was going to persevere in it, they came in throngs from the countryside and the town to present themselves for apprenticeships." With an ample supply of young workers on hand, the satisfied producer alerted his confrère in Ambert that he would steer any promising youths his way.[11]

Much depended on the aspirant's links to Vidalon-le-Haut. Ambroise

Guignal landed his apprenticeship on the word of a priest.[12] A reliable advocate at the mill, in this case the governor Pacquet, cleared the way for Jean Bacon's indenture.[13] Above all, the Montgolfiers counted on the information they learned about the candidate's family. Did the young man have roots in the craft? The son of a stonecutter, Jean Texier won a spot as an apprentice: surely his father's innocence of papermaking and the "modes" enhanced Texier's prospects.[14] Was the candidate's family intact? That is, could the Montgolfiers turn to the boy's kin for aid in disciplining him? Barthélemy Géry's mother had died, but his father, a day worker, lived in nearby Serrières.[15] Orphans were in particular need of the blessing of someone trusted by Vidalon's *patrons*: the backing of one of the Montgolfiers' cousins got Mayol his apron.[16]

Family credentials, however, revealed little about an aspirant's aptitude or diligence. To gauge these intangibles, the Montgolfiers' inspected the newcomer's body, speculated about his (or her) vigor, and resorted to physiognomy. Joseph Bourgogne was robust; Baugi was "small but well-made"; Antoine Voulouzan "of good size for his age [and] sufficiently well-proportioned."[17] A solid constitution clearly impressed Vidalon's *patrons* more than precocious size—the Montgolfiers suspected that Pourra's rapid growth at an early age had left her weak and "a little nonchalant."[18] A "rather handsome face" enlivened one youth's chances, since the Montgolfiers read much from a youngster's physiognomy.[19] Texier had a skillful "air"; two brothers from La Lombardière gave the impression that they could get things done. Tainted by a "shameful demeanor," poor François Perrin landed a place but did not last long at the mill. Vidalon's *patrons* were never sure what to make of François Cordier: his mien said nothing. Lucky André Force offered both a "pleasant face" and a "good body," and strong Jean Barjon, with his determined countenance, was just the sort of lad who might stand up to the journeymen of the paperworkers' association.[20]

Dependent on physiognomies and pedigrees, the Montgolfiers were careful about making commitments to the youngsters who sought work at Vidalon-le-Haut. They engaged Jean-Pierre Charat as an apprentice in April 1787 and agreed to arrange terms with his father if the métier suited the boy. Eleven months later, Jean-Pierre and Charat père reached formal accord.[21] Before endorsing apprenticeship contracts, Jean-Pierre often put promising youths through probationary periods: Girard, for instance, was hired "on a trial basis as an apprentice."[22] Sometimes such auditions ended quickly. Eighteen days after Jean-Pierre engaged thirteen-year-old Louis-Joseph Brunel, he lambasted him as "a child accustomed to doing nothing who has a great deal of trouble getting down to work."[23]

Although the Montgolfiers continued to train apprentices throughout the 1780s, seasonal surges in production and the discharge of unruly hands also compelled them to hire journeymen. Engaging a veteran like Louis Barré promised few surprises: he came to Vidalon-le-Haut from Augustin Montgolfier's mill in Savoy, which was directed by a former Vidalon formaire. Even so, Vidalon's *patrons* cautiously agreed to set Barré's wages after they had seen his work.[24] After the lockout, the Montgolfiers dealt warily with men who had learned the craft from other papermakers, even when they knew these producers well. To assess a journeyman's potential, they demanded to see his *certificat de congé*, the discharge papers endorsed by his previous master.

In 1671, the state barred papermakers from engaging any wandering hand unless he could produce a written certificate of dismissal. This measure emerged from the lawmakers' belief that journeymen in all trades were inherently turbulent and that men on the move were particularly troublesome. Constraining the paperworkers' wanderlust, a concern that reappeared in royal decrees across the eighteenth century, would curb their excesses—or, to put it another way, prevent the journeymen from refreshing and reinforcing their sense of the proper order of the trade. As one Auvergnat official declared, the ease with which "seditious" paperworkers quit their bosses and secured jobs elsewhere was "the abuse whose reform would contribute infinitely to suppress the other abuses."[25]

But this reform introduced its own abuses. A spiteful master might paint an unjust picture of a journeyman's habits or skill, exaggerate an outstanding debt or even fabricate an obligation owed by a hapless worker, or simply withhold the document altogether. "Always bearers of several *congés*," paperworkers found their own means to circumvent the law.[26] They also trafficked in counterfeit certificats, two of which permitted a pair of journeymen who had taken sudden leave of their boss in Chamalières to land work in Thiers in 1788.[27]

Certainly, fraudulent credentials bedeviled Jean-Pierre Montgolfier, who did most of Vidalon's hiring and firing during the 1780s. In an era when papermakers commonly poached hands from one another, he sacked the governor Le Bon and his wife when he discovered that they had won his confidence with tainted references. The pair had abandoned Thollet's mill in a nearby hamlet "without his consent" and tricked his son into providing them with honorable discharges. Once Thollet père alerted Jean-Pierre about the ruse, the latter sent the governor and his wife packing, with a plea to return to their former master. Although Jean-Pierre had complied with the law, he admitted ruefully that he lacked a sample of the younger Thollet's handwriting.[28]

Generally, however, paperworkers could obtain a wage and lodging without written references. In fact, Thollet's mill welcomed at least two other Vidalon hands who had slipped away without properly taking their leave.[29] Some journeymen evidently put little stock in their own credentials: Claude Durif and Pierre Joubert fled Vidalon-le-Haut noiselessly, with their certificats de congé still in Jean-Pierre's hands.[30] Yet, the veteran who failed to take proper leave of his former master soon learned that he had little chance of winning a job at Vidalon-le-Haut. Of the fifty-eight journeymen paperworkers and governors secured by the Montgolfiers from April 1784 through 1789, at least forty-six (79.3 percent) carried a written discharge. Two of the remaining twelve had letters from Augustin Montgolfier, a third had the word of Antoine-François Montgolfier of Vidalon-le-Bas, two more came from a mill close at hand, and one, Jean-Pierre emphasized, had been hired in his absence.[31] Typically, Jean-Pierre would not consent to Duranton's request for work for himself and his family in 1784 without the approval of his current master, Antoine-François Montgolfier.[32]

For most papermakers, who tolerated the "modes" in order to fill out a vat crew for a few days or even a short production season, a glowing congé provided little assurance and provoked limited interest. But the Montgolfiers at Rives and at Vidalon-le-Haut were building work forces of malleable hands for the years. The journeymen of both Thiers and Dauphiné bitterly resented Augustin's attention to the document and made his former hands pay for the right to labor in their company; the fine was 10 livres.[33] Occasionally, Vidalon's bosses may have also paid for their resolve. On April 9, 1787, they hired Etienne Lebord, who possessed a discharge in good order. He disappeared on the morning of the tenth, "without saying anything." Had Lebord traded on the Montgolfiers' widely known commitment to the document to gain a free night's shelter?[34] Trusting Grenier's and Sauvade's discharges certainly cost Jean-Pierre. Their papers portrayed them as full-fledged journeymen, but once he saw their work, Jean-Pierre muttered that they were no better than his apprentices. Eight days after he noted the deficiencies of the two, Jean-Pierre gave Sauvade six weeks' notice. He was gone three days later.[35]

Jean-Pierre was especially cautious when a journeyman had taken his time on the road to Vidalon-le-Haut: had he hidden stints in shops where the "modes" prevailed? Accordingly, itinerants had to account for both their stops and their means of survival along the way to earn a place with the Montgolfiers. Durif told Jean-Pierre that he had not worked during the four months since he had quit a mill in Dauphiné.[36] Claude Artaud made the same claim about three weeks in 1785.[37] A discharge dated at the end of May 1785 cleared the path for Benoît Chanteloube to secure a job at Vidalon-le-Haut in August.

Chanteloube's candor surely didn't hurt either: he confessed that he had toiled for a few days elsewhere "in order to make his way."[38] Caught in the same snare, the veteran Pierre Chêne acknowledged that he had put in some unrecorded days at another mill.[39] And Joseph Russet, who had journeyed to Annonay from distant Savoy, where the "modes" remained in full force, was greeted with a long set of mill ordinances.[40]

When the Montgolfiers fancied a journeyman's skills or presence, they tied him down even if they lacked an opening around the vats. They provided him with a temporary wage as a common laborer, perhaps on their lands. Thus one tramp with proper credentials began his career at Vidalon-le-Haut as a "*manoeuvre* on account of the drought."[41] When necessary, Jean-Pierre also took on a journeyman or two on an explicit, short-term basis. Pons, for one, got two days of work while some regular Vidalon hands recovered from illness.[42] Most often, however, the Montgolfiers engaged journeymen for ordinary, full-time service. They hedged their choices by mandating trial work periods for eighteen of the fifty-eight journeymen (31 percent) mentioned above. (Eight more of these veterans escaped probation but had to display their skills before the Montgolfiers settled their wages.)[43] The probationary period was occasionally open-ended, but it usually extended no more than two or three weeks. Typically, Michel Duranton put in twenty days before Jean-Pierre hired him on a permanent footing.[44] Trial engagements gave the Montgolfiers time to evaluate workers from distant mills and allowed experienced hands a grace period to determine if Vidalon's nouvel ordre was for them. When a paperworker from Normandy arrived at Vidalon-le-Haut with an endorsement from Besançon, the Montgolfiers hired him for a fortnight to see if they were compatible.[45] Other considerations led Jean-Pierre to engage Antoine Duranton on probation. A disabled ex-soldier, Duranton appeared at the mill with a certificat de congé signed by his brother. Jean-Pierre refused to set his wage until he had seen Duranton's work.[46]

The Montgolfiers' new shops required a swarm of female hands to sort, shuffle, and count. They trained many themselves but also depended on the wives and daughters of journeymen. These women did not participate directly in the customary lives of their men, but the Montgolfiers feared that they were all too familiar with their rights and rituals. So Vidalon's *patrons* demanded that women on the tramp also carry certificats de congé. Traditionally, however, master papermakers apparently did not issue written dismissals to *ouvrières papetières*.[47] It was not until the 1780s that the Controller General Joly de Fleury ruled that papermaking women must carry discharge papers, "the word 'worker' [in the edict of 1739] being generic for both sexes."[48] This mandate dovetailed perfectly with the Montgolfiers' desires.

Grenouillet's wife landed her spot at Vidalon-le-Haut with discharge papers provided by Antoine-François Montgolfier.[49] A written endorsement by the Johannots (as well as the recommendation of a local curé) opened a job for Claire Beraud.[50] Even so, Vidalon's women workers, like Marianne Joubert, who began her stint with the Montgolfiers in September 1782, often served probationary periods.[51]

Anne Chanteloube's return to Vidalon-le-Haut in January 1787 revealed the Montgolfiers' adherence to the royal decrees that suited them, regard for proprieties between masters, and concern for the orderly conduct of workers—even women workers. Chanteloube had left the mill several months earlier to toil as a *fille papetière* in La Saône. Although the Montgolfiers knew her well, they refused her request for work: she did not possess a discharge endorsed by her last master. Her mother claimed to have written to La Saône to secure this document, but the wary Montgolfiers suspected that she had done nothing of the sort. "These people," Vidalon's *patrons* commiserated, "are poor ones who have need of their work in order to survive." Thus the Montgolfiers themselves addressed a letter to La Saône to discover if Chanteloube's former boss wanted her back. If so, they pledged to return her; should the Montgolfiers fail to receive a response within eight days, they promised to hire her. Still, they concluded with a vow to ship her back to their confrère "any time that you ask for her during this current year." Neither Chanteloube's own desires nor the Montgolfiers' need of her services would prevent them from dispatching her to La Saône. Even more remarkably, they expected Chanteloube's mother to aid them in making the state's regulations work; that is, they expected her complicity in the policing of her own daughter's journeys.[52]

Whether hired on a trial footing or not, every journeyman and most apprentices received a number in the Montgolfiers' account books, a practice Vidalon's bosses had followed long before the lockout. (Since fathers were paid for the work of their sons, they often shared a single number.) The Montgolfiers also issued a small record book (*carnet*) to their journeymen and apprentices. These booklets provided the workers with running accounts of the advances and regular payments that the Montgolfiers had entered in their own ledgers. There was no question who controlled the carnets: in November 1785, Jean-Pierre reported that a payment made by his brother had not been logged in Sauvade's booklet; Jean-Pierre closed the matter by making the entry himself.[53]

Despite a charge for each new carnet, some workers repeatedly lost them. The governor Chirol left his in town, and in 1788, Michel Chantier had to fork over 18 deniers for his third carnet.[54] These little logs, not surprisingly, had a

habit of disappearing at just the time when a worker received an advance or his routine pay. Jean Frappa's receipt of 12 livres in November 1785 was "not written in his book which he did not have at the moment." Unfortunately for Frappa, Jean-Pierre concluded his account of this transaction with a note that it had been marked in the worker's carnet.[55]

To remain on the Montgolfiers' payroll, Vidalon's hands had to observe the mill ordinances inscribed in their carnets. Thus reward and discipline were linked every payday. These regulations, the *usages de la maison*, welcomed each newcomer to the mill. Accordingly, the first lesson the Montgolfiers instilled in the apprentice Gonein was that he must obey "the *usages de la maison* as they are expressed in his [record] book."[56] In October 1786, one fresh arrival pledged "to abide by" the house codes and a second "promised to observe" them.[57] Reflecting the Montgolfiers' vigilant posture after the lockout, these rules prohibited interference by workers in the affairs of their fellows, fines levied by journeymen, and all of the "modes," particularly drinking in the company of the transient rentes. Put succinctly, the "customs of the house" outlawed the journeymen's custom: "I do not suffer any *mode*," Jean-Pierre declared in 1784.[58]

The usages de la maison taught the newcomer the Montgolfiers' formula for harmonious shops: do what the bosses want, when they want it, with whom they select, with the proper attitude. As Joseph Buisson discovered, he was to toil at the job assigned to him and to do so beside the men chosen by the Montgolfiers. He was also never to drink around the vats.[59] Violet and his sons learned that they would retire on Sundays and holidays "at eight o'clock at the latest."[60] Even before the journeyman Tissier received a day's pay, he found out that abandoning his work without permission would cost him 40 sous. Moreover, his new masters demanded that he behave "with decency."[61] To continue his apprenticeship at Vidalon-le-Haut, Charat must remain "tranquil"; to please his new employers, Russet should be "submissive and obedient."[62] In 1785, the word "modes," though not the ban on worker involvement in the business of their mates, disappeared from Vidalon's house customs. Here was a measure of a certain confidence as well as persistent caution.

When a worker left Vidalon-le-Haut, he was caught up in a new round of paperwork. The Montgolfiers updated his certificat de congé, assessed penalties for sick days and drinking bouts, appraised his attitudes and skills, and determined whether—and in what circumstances—he or she should be rehired. Many of their evaluations were terse yet sent a clear message: Marie Rivoire was "without reproach"; Charbellet was "a very bad subject."[63] "Hav-

ing nothing against him," Jean-Pierre Montgolfier provided Poynard with an honorable congé.[64] Yet, in fits of anger and delight, Jean-Pierre occasionally went on at length. This characterology, of course, served a purpose. Because many of the Montgolfiers' former hands returned time and again, Jean-Pierre could consult something more than facial shapes and family trees. Jelbi, a "well-behaved lad, not debauched, understanding reason but a bit fickle" was the sort to rehire "with haste anytime."[65] Millot was the type to take back after he had knocked around and "learned manners."[66] Marie Valençon, a clean woman who worked well enough, was "a bit headstrong" and hence merited only a lukewarm endorsement.[67] Durif and Joubert should only be given work in a pinch, "without counting much on them."[68] Still, Jean-Pierre knew that he was evaluating fledglings and frequently hedged his bets. About Fagot he wrote: "young; [I] don't know what this one can become."[69] Jean Chantier performed his tasks well enough but liked his wine: would he give up his ways or give himself up to *"la débauche"*?[70]

Like other Old Regime employers, the Montgolfiers attracted workers with advances. When stints ended badly or abruptly, Jean-Pierre had no choice other than to note these uncompensated funds in the departing worker's congé. Since Jacques Roux quit the Montgolfiers with an outstanding debt of 16 livres 1 sou 3 deniers, Jean-Pierre wanted it marked as "pressing" in his written discharge.[71] When Bombru, who owed Vidalon's *patrons* the unusually large sum of 75 livres 15 sous 6 deniers, asked to be released from his indentures, Jean-Pierre obliged.[72] His debt must have been depicted as pressing too, since his wife remitted 27 livres to the Montgolfiers two years later.[73] For François Jamet, a Montgolfier élève who dared to make his tour de France, a cleared debt was the price for his certificat de congé.[74]

In May 1785, an exasperated Jean-Pierre wrote to Thollet, whose mill was a favored refuge for Montgolfier hands who decamped *sans certificat*. Antoine Gagneres, a Vidalon carpenter, had left without warning and, even worse, had slipped 12 livres of Montgolfier advances into his pocket. Jean-Pierre learned (from a Vidalon hand?) where Gagneres had landed and did not wish to pressure him to return. He did, however, want the 12 livres back and demanded that Thollet deduct this sum from his future wages. Jean-Pierre also requested an immediate response from his confrère, observing that he would have to reconsider his course should Thollet reject his appeal. He was certain of one thing: he did not intend to be Gagneres' "dupe."[75]

Jean-Pierre pursued Augustin Valençon, yet another man who fled to Thollet's shops, with particular vigor. His debt to Vidalon's *patrons* was 54 livres, an especially substantial sum for a young man who was only an apprentice beginning the layman's craft. Jean-Pierre had advanced the money to Valençon

for clothing, and counted on recouping his investment in the youth when he was skilled enough to earn regular wages and premiums. The Montgolfiers had every legal right to reclaim this runaway novice, but they were willing to settle for repayment of the 54 livres. Again, they asked Thollet for the funds; with that accomplished, they would permit Valençon to remain in his employ. Meanwhile, Jean-Pierre betrayed his frustration with ungrateful workers and a thin skin about their capacity to gull him: "I don't wish to be the dupe of my workers anymore, who have always paid with ingratitude for the services I have rendered them."[76]

Two weeks later, Jean-Pierre wrote again to Thollet. He considered Valençon's brusque departure "a loss beyond the money" he had loaned him. "After having learned his craft at the expense of my merchandise," Valençon had reached the point where he could render valuable service. Still, if reimbursement had been assured, Jean-Pierre continued, he would permit this useful youngster to ply his trade "tranquilly" in Thollet's mill. But Valençon had refused to pay up. Thus Jean-Pierre concluded: "I believe that it is advantageous for you, for me, and for our *confrères* to make an example." Return Valençon to me, he demanded, "for having come without [a] *certificat* and without having satisfied his debt to his first master."[77]

When all else failed, Jean-Pierre resorted to the blacklist. Louis Pichat's apprenticeship with the Montgolfiers ended when they caught him making a fire with their wood. Jean-Pierre subtracted 16 livres from the youth's outstanding wages and, with that settled, noted that he would not obstruct any producer who wished to give Pichat work.[78] Although he had been sacked, Pichat might have considered himself fortunate; at least Jean-Pierre did not blacklist him. An angry papermaker could cashier a journeyman without providing a written discharge, as Jean-Pierre did when Blaise Duranton "mutinied."[79] He could thwart a worker's chances of finding a job elsewhere by portraying him as an "incorrigible drunkard," as the vatman Joseph Etienne quickly learned.[80] How helpful, finally, was Millot's discharge? While affirming his skill around the vat, Jean-Pierre warned that he had been unable "to submit to the *règles de la maison* [mill rules]" and had struck a fellow worker.[81]

On April 30, 1781, Jean-Pierre hired Barthélemy Dausson, a paperworker who carried an endorsement from Vidalon-le-Bas. On May 31, he gave Dausson six weeks' notice.[82] Whether the finish was stormy or silent, it rarely lasted the forty-two days mandated by the government. Which papermaker, after all, wanted his shops troubled or his work sabotaged for a month and a half by an unhappy vatman? Despite his self-proclaimed loyalty to the state's edicts, Jean-Pierre abridged the leave-taking period when it served the Montgolfiers'

interests. He canceled the obligation for one female hand "out of consideration for her, [a] good worker, [a] mild and quiet woman." On the same day, he released Duranton's wife from the requirement "in order to be rid of her." She was neither clean nor orderly, spent too much time caring for her children, and produced little work.[83] Workers, too, found reasons for quick endings. After two days with the Montgolfiers, one woman quit to sell wine in Annonay.[84] Jean Deschaux stalked out of the mill in October 1786 with some pay and his certificat de congé in Jean-Pierre's hands. He was angered by the frailty of Vidalon's apprentices; perhaps he was also implying that he did not intend to labor beside young men who had not been tempered by the "modes."[85]

Occasionally, workers remained at Vidalon-le-Haut for the full six weeks after giving notice. Louis Vallet, for instance, made his decision on February 4, 1787, and left the mill on March 17.[86] When workers stayed on after serving notice, some sort of negotiation or reconsideration was generally afoot. On July 2, 1785, Payan's wife, evidently following her husband's lead, advised the Montgolfiers of her determination to move on. When Payan withdrew his resignation on the third, she followed suit on the fourth.[87] The journeyman Périgord apparently voiced his desire to depart as the first step in a round of bargaining. Three days later, with the promise of an annual bonus in his pocket, Périgord agreed to stay.[88]

In 1781, Versailles decreed that the journeymen in all trades must join their discharge papers in a small notebook, the livret.[89] Moreover, the state reduced the leave-taking period to eight days, a realistic reform. Because the livret, which was to function like a domestic work passport, would remain in the employer's desk, fewer journeymen would presumably feel free to abandon their bosses without notice. The reliable historian of Auvergnat papermaking, Henri Gazel, insisted that the *lettres patentes* of 1781 were neither enforced by the state nor obeyed by the papermaking industry.[90] Yet, the Auvergnat journeyman Jean Peschi arrived in Annonay in 1786 with references from Troyes "in conformity with the *lettres patentes* of September 12, 1781."[91] Jean-Pierre, then, was aware of the revised form. Like the rest of his trade, however, he spoke only of the customary six weeks' notice and honored it most often in the breach.

The Montgolfiers looked for journeymen with clean slates and for solid, self-confident novices with transparent faces. Hiding nothing, preserving nothing, unattached to the "modes" and the paperworkers' association, these men would be open to a different kind of initiation. They would be employees, separated from the floating swarm of paperworkers at the moment of hiring by pledges and a paper trail of the Montgolfiers' and the state's making. Not every element of Vidalon's hiring and firing procedures after the lockout

was new, but these practices took on fresh meaning and urgency within the Montgolfiers' nouvel ordre. Such adaptability and flexibility remind us that Old Regime institutions and forms were anything but closed to novel managerial imperatives. Thus, sometime after their former journeymen had served notice en masse, the Montgolfiers rechristened their *"lit des rentes,"* the room that sheltered these tramping men, the *"lit des ouvriers."*

CHAPTER NINE

Paternalism

Pierre Montgolfier, Vidalon's patriarch, despised the workers' "modes." These customs, he knew, nurtured the journeymen's independence and sustained their loyalty to fellow workers and familiar techniques. He considered all this an offense against the proper order of his trade and, even more, contrary to the eternal hierarchy of rule and submission. Thus he believed that master papermakers must also fulfill enduring duties. He worried about the "abuse" that could result from "the excessive authority of certain masters."[1] His task was to transform Vidalon-le-Haut into a moral community, in which technological advance, efficient production, and catechized novices stood apart from the tramping journeymen and their self-destructive ways. He agreed with an anonymous mémoire that the "modes" left the workers "perpetually miserable and ruin[ed] their health."[2] Instead, Pierre proposed that his hands attend to "everything that affects their master's interest" in exchange for food, wages, and shelter.[3] A second code spelled out precisely the sort of behavior the Montgolfiers intended to encourage. Since the master "cherish[ed]" his workers as his children, it was just that they acted on his behalf. Consequently, all the men who sweated in each of his shops would be liable for thefts or damage until they named the culprits.[4] Whether the Montgolfiers offered protection, provisions, advances, or solicitude, they exacted a high price for their concern: the frustration of any solidarity among their new hands.

Vidalon-le-Haut provided workers and their families with a place to live, worship, rear their children, and, of course, work. More than a set of workshops, the mill had many trappings of a village.[5] Austere, authoritarian Pierre Montgolfier was cut to govern this village with a firm hand. The paperworkers' combination, an instrument of mutual aid and self-sufficiency, had stayed that hand. The lockout, however, furnished Pierre and his sons with the chance to substitute vertical ties for horizontal bonds. Such dominion may conjure up friendly, even warm, images, but the paternalistic tradition that Pierre passed to his children was tough-minded and clear-eyed, as carefully calculated as wage incentives or the coucher's daily responsibilities. It was a matter of exchanges, a sort of transaction, however unequal, between manu-

facturers and the skilled, or those just mastering the trade. It was, nonetheless, tempered by a certain intimacy that allowed the Montgolfiers to minister to their workers' individual needs, the better to win their confidence, perhaps even their allegiance.

Paternalism, then, was an integral element of the Montgolfiers' new order. It was not a step on the path to rational, impersonal management but its essential complement. In return for the responsible care of Vidalon's tools, one of the Montgolfiers pledged to sustain his workers with "all the comforts that are in his power."[6] Here were incentives to tether young hands to the mill, incentives that promised enough time for Vidalon's *patrons* to transform bons élèves into paperworkers.

Already freighted with disciplinary expectations, the material benefits of a stint at Vidalon-le-Haut also entailed monetary charges. Grenier's annual rent in 1785 for one of Vidalon's *"petites chambres"* was 12 livres.[7] Other quarters, which were strewn among the workshops, went for 10 livres. For comparison, the yearly rent for a chambre in Bayeux was 10 to 20 livres, with higher charges common in only the most populous cities.[8] There were thirty-two worker apartments at Vidalon-le-Haut. (The Montgolfiers also resided in the manufacture, but their quarters were hived off from the workers' flats and the ateliers.) Some rooms contained a single bed; one space, apparently designated for unmarried male apprentices, housed seven. Each compartment included a fireplace, wooden beds with straw mattresses, and bedsheets, whose linen was rougher than that used by the Montgolfiers.[9] Still, these quarters were probably superior to those reserved for unmarried men in a typical paper mill: "Nothing was more distressing than to see their pallets heaped up on the soil, in dirty, humid *chambres* almost without daylight and air."[10]

Worker families at Vidalon-le-Haut could expect a room to themselves, but Jean-Pierre occasionally housed women with absent husbands and unmarried ouvrières in these family quarters. The promise of a room of their own was evidently the bait that lured some worker families to the mill. When the journeyman Filhat and his family arrived in Annonay, Jean-Pierre Montgolfier agreed that they could live rent-free until they enjoyed the exclusive use of a chambre.[11] Moreover, Jean-Pierre, who pinched every sou, remitted the rent of one worker and his wife; they "did not have a *chambre* to themselves" during their brief stay.[12]

Exhausted and "nearly nude" when he reached Vidalon-le-Haut, Vincent, who became a servant, needed a complete wardrobe: a shirt, a hat, a pair of sabots, two pairs of culottes, a jacket fitted with new sleeves, and a half-worn pair of shoes. The Montgolfiers docked him 17 livres for this ensemble.[13] Vidalon's *patrons* provided frequent advances in cash and in kind to jour-

neymen and apprentices. They lent 12 livres to the apprentice Charat for clothing and paid a local woman the same amount to outfit Jean Texier when he returned to their employ.[14] In October 1779, Marguerite Savoye reimbursed the Montgolfiers for "the past month's coal."[15] Some workers grew quite accustomed to these advances. One borrowed 6 livres for apparel in September 1784, 21 livres to fetch his wife and purchase a dozen shirts in September 1787, and 14 livres in March 1788.[16] Several hands even built debts before they began work. The apprentice Quinat, for instance, agreed to repay an advance of 12 livres from his first wages.[17]

Troubled workers also turned to the Montgolfiers for small loans: in April 1783, three hands borrowed 4 livres 10 sous apiece to get out of jail.[18] Of course, the Montgolfiers kept careful records of these advances. When the journeyman Jean-Joseph Micolon died, Jean-Pierre dutifully recorded that he had to write off the funeral expenses.[19] Entangling a large number of workers in a web of debt, however, did not suit the Montgolfiers. They were sorting through the sons of diverse craftsmen, day workers, and vineyard hands in search of paperworkers. While the Montgolfiers extended limited credit and welcomed the dependence it produced, they valued their prerogatives and well-ordered shops above reimbursement.[20] Jean-Pierre did not let the likely loss of headstrong Blaise Duranton's obligation of almost 10 livres stand in the way of his departure.[21] In fact, the debts Vidalon's *patrons* pursued most vigorously were those owed by workers who had fled the mill sans certificat.

Expelling the journeymen's "illicit" custom from the mill did not prevent the Montgolfiers from blending the trade's "licit" customs (those that served the bosses' ends) into their regime. Once a fixture in the craft, the master's table was increasingly rare in the paper mills of the Vivarais. "In these cantons" Pierre Montgolfier alone continued to feed his workers, claimed one of his sons, although it was a "surcharge" on his mill.[22] Pierre, however, believed that the provision of food and housing brought him "choice" hands and ushered more families of workers into "the shadow of the craft."[23] The Montgolfiers themselves continued to eat with their workers, or at least the male hands, as was customary in the small mills of their Auvergnat roots.[24] Yet, the appearance of liqueurs, almonds, coffee, and chocolate in their accounts—items clearly not intended for workers—suggests that they took some meals apart.[25] Meanwhile, the mill women probably abandoned the master's table to their men, as did their counterparts in the Auvergne.[26] The Montgolfiers' female hands received an allowance of 5 sous per day for food, as well as soup to pour over their bread.[27]

The Montgolfiers' assessment of the general decline of the master's table in French papermaking seems accurate. In 1751, the proprietors, tenants, and

journeymen of the royal manufactory of paper in Angoumois reached a collective agreement. Despite the elevated title of the linked mills that formed the manufacture royale, the local industry had fallen on hard times. During the previous eighty years, more than sixty mills had slipped into "total ruin."[28] To ensure the survival of the remaining enterprises, masters and men came to an elaborate accord about hours, output, recruitment, and above all, nourishment. Ritual and symbolic capital figured in the bargaining, but largely as adjuncts to meat-and-potatoes issues. In exchange for "fat wages," the bosses would no longer feed the journeymen.[29] The traditional banquets that celebrated a move up the ladder of skill, an apprentice's arrival at the journeymen's table, or the death of a paperworker were reduced to toasts. Rather than the upwardly mobile apprentice or journeyman contributing the princely sum of 150 livres to the workers' treasury, he would now offer the meager fee of 10 livres—or so the manufacturers wished to believe.[30]

A papermaker named Trenty, proprietor of two mills in Agenais, wanted to imitate his confrères in Angoumois. He realized that liberating himself from the obligation to nourish his workers was no easy thing, so he sought the support of the provincial intendant. A report was drawn up to address Trenty's proposed action, which noted that "the number and the quality of the dishes are fixed and assigned for every day and for every meal." Moreover, the journeymen's "etiquette" specified how the boss and their brother journeymen "must receive them, treat them, greet them, and even face them." Should one of these "syllables" go unsaid, all work ceased, the workers deliberated—"God knows how," the mémoire observed—and penalties were pronounced. Small wonder that Trenty wanted to put an end to this flashpoint.[31]

In 1765, Pierre Montgolfier and his son Raymond maintained that their workers "were fed at the expense of the *patron,* and generally well nourished." With picturesque flair, father and son observed that these men "lay the tablecloth three times a day." In the morning, the Montgolfiers provided soup, a piece of butcher's meat, and some streaky bacon or pickled pork. The midday meal included soup, some fricasseed vegetables, such as peas, beans, or potatoes, and Gruyère cheese. Supper reproduced the morning's fare, a meal worth 1 livre per man, according to the Montgolfiers. Wine, diluted to "have peace in the household," accompanied each meal.[32] (It was also held that wine inhibited the diseases endemic to the trade, from varicose veins and edema to rheumatism.)[33] A quart and one third of unadulterated wine, known as a *pot*, and bread composed the late afternoon *collation* that rewarded "extraordinary" labor.[34]

The workers found this array less than appetizing. They complained that the bacon they received for supper on Sunday, Monday, Tuesday, and Thurs-

day was "the scrap [*débris*] from the sack." Prepared badly, it was inedible. The wine the Montgolfiers offered was "the worst wine you can find," the workers grumbled, "because very often it gives us a lot of diarrhea." "Everything" that Vidalon's *patrons* provided, claimed the workers, was "contrary to our health." Perhaps the master's table was less of a surcharge than the masters let on. Evidently, it also failed to nourish the bonds the Montgolfiers desired. On the basis of their meals, the workers concluded that "Monsieur" was "not happy with us." The workers had caught on to the Montgolfiers' calculated paternalism and turned the tables: "You insist that your work be made and we require that the food you give us be better prepared than it is."[35] Every disciplinary regime, it is widely known, provokes its own forms of resistance; so does bad food.

The workers were not alone in grousing about the master's table. Pierre and Raymond complained bitterly about holiday meals that "custom authorizes" and that their hands demanded "with arrogance."[36] The Montgolfiers' outrage should be taken with a grain of salt, however, since they charged for these holiday delicacies: in 1784, the worker Marolle's contract stipulated an annual wage of 60 livres when he sat at the master's table on Sundays and feast days, and 90 livres per year when he did not.[37] Across town, a pact reached in 1770 by the Johannots and their hands specified that the workers would put in a quarter day of work on Good Friday in return for their usual food, wage, and daily "bonus." The Johannots also agreed to grant the customary pot of wine to each vat crew that fashioned the unwieldy *grands papiers*, but only after the team had attained its quota. Like the Montgolfiers, the Johannots retained those "licit" ways of the craft that meshed neatly with their disciplinary needs.[38]

Interestingly, paperworkers also rushed to the defense of the trade's "licit" custom when it served their own interests (and bellies). A threat to the master's table in 1541 had prompted the Lyonnais paperworkers to argue that "if they were going home or to the tavern to eat, it would hardly be possible for them to find themselves together in the shop at the same hour." The result would be idle hands and fewer reams: when "one of them is absent," they concluded slyly, "the others can't begin work."[39]

Shrove Thursday (*jeudi gras*) festivities apparently reminded Vidalon's *patrons* of their traditional obligation to nourish all their workers. In 1785, the Montgolfiers celebrated this feast day with a collective gift of 28 livres to their female hands.[40] A year later, they provided bounties to the women and the "men who feed themselves."[41] The latter offering reflected a new option for Vidalon's male hands. These men were now "free" to take their meals at the master's table or to receive a cash allowance for their food.[42] Perhaps Vidalon's

bosses provided this choice in order to trim troublesome hands from their table; after all, successive mill codes warned workers to avoid swearing or shouting at the table, and to retire to their rooms rather than appear at meals overcome by wine.[43] Sixteen male workers, roughly a quarter of the Montgolfiers' mature male work force, elected to abandon the master's table in December 1785. Among them were seven long-term apprentices, in effect bound journeymen who had been at the mill long enough to perform skilled tasks around the vats.[44]

Whatever moved the sixteen to quit the Montgolfiers' table, it was probably something more than the size of the stipend. Vidalon's *patrons* offered 16 livres 10 sous per month, a figure identical to the value they placed on each man's portion in their dining room.[45] Perhaps there was merit in the workers' complaints about the food served by the Montgolfiers; perhaps these men simply wanted to dine far from the bosses' watchful eyes. Once it had been an honor for young workers to join the company of veterans around the table, commemorated by one of the "modes." But under their new regime, the Montgolfiers had to urge the men who remained at their table to hurry to their meals when the saleran called.[46] In the absence of the laughter and esteem contributed by the journeymen's ways, many of Vidalon's fresh hands failed to develop much loyalty to the master's table.

Vidalon's *patrons* would not bear the expense of feeding incapacitated workers. Certainly, illness spread easily in their mill. In a subterranean chamber, women huddled together on benches sorting filthy rags; in the clammy heart of the fabrique, the journeyman dipped his mold into a vat of heated water several thousand times a day; and in the drying loft, women were constantly exposed to drafts. Congested living arrangements and the closeness of the master's table added to the toll. If a worker intended to linger in his room "due to illness or other causes," he had to notify the Montgolfiers of his plans before breakfast or pay for the entire day's food.[47] On March 13, 1788, the worker Thibert evidently did just that, since Jean-Pierre deducted only his lost wages.[48] Of course, Jean-Pierre was determined to be nobody's fool. When the convalescent Jean Texier joined François Charbellet in leading three other apprentices on a spree, Jean-Pierre was furious. He had kept Texier at the mill during two months of sickness and the beginning of his recovery; now he "drove [Texier] away."[49]

Infirm workers often headed home to heal within the orbit of their families and to avoid charges for food and rent. Nesmes remained with an uncle for seven weeks when he was ill.[50] Jacques Roux, who hurt his leg while playing in the mill, spent three weeks at home.[51] Chaumier, who missed two months in the autumn of 1784, recovered with his parents and in the *hôpital*.[52] When

necessary, Vidalon's *patrons* made arrangements for the hospitalization of their hands, as they did for Grenier in April 1784.[53] Whether a worker was home or in the hôpital, such leaves also fulfilled the Montgolfiers' needs, freeing beds for productive workers. Furthermore, the burden of sustaining a worker while he regained his strength or of ferreting out a malingerer was lifted from Jean-Pierre's shoulders.

Louis Gélimau was "almost always ill."[54] The Montgolfiers knew which of their workers were plagued by poor health and were especially wary of them. More than one journeyman had been *"toujours malade,"* produced little paper, and eaten the bosses' food during a brief stint at Vidalon-le-Haut. Recently recovered hands were also suspect. After a three-month spell in the hôpital with fevers, Antoine Joubert secured a place in the mill, Jean-Pierre noted, "in my absence."[55] Experience had taught the Montgolfiers to link illness and unreliability; on the departure of the ailing worker Pourra, Jean-Pierre commented that "he was subject to frequent sickness and became debauched."[56] The Montgolfiers dealt unsentimentally with sick workers; they were too mindful of lost work to act otherwise.

Making paper by hand was family work. Even the smallest mills provided employment for wives and older daughters, as well as odd jobs for younger children. Marie Chatron, for instance, toiled in Vidalon's sizing room when she was eight years old.[57] The Montgolfiers believed that child labor was particularly beneficial: it fattened the family purse while implanting the "habit of work." Too many children, however, could sap the energy of an industrious hand. Thus Jean-Pierre fretted that Michel Duranton's wife, a "worthy" woman who performed her tasks well, was "burdened with a large family."[58] Still, a "nursery of useful apprentices," as Augustin depicted it, meant less damage in the mill while "little by little" the youngsters picked up the trade.[59]

The Montgolfiers made use of their workers' families to build and police their internal labor market. They depended on fathers, mothers, brothers, uncles, cousins, and in-laws as recruiters. They also counted on senior family members as aids in disciplining recalcitrant youngsters. For the workers, family connections provided job opportunities, smoothed over rough spots with the *patrons*, and offered a favored pretext, in the form of a trip home to arrange family business, for those hands who wished to vanish. Accordingly, the Montgolfiers' reliance on worker families rendered them vulnerable to the families' interests, whether it was a matter of placing a dissolute youth or papering over a departure sans certificat.

Worker families commonly arrived at Vidalon-le-Haut in search of work. In December 1785, Filhat, his wife, and their three children presented them-

selves to Jean-Pierre. Sixteen months later, another Filhat and his wife secured employment at the mill.[60] But journeymen and their spouses did not always reach Vidalon-le-Haut in tandem. In 1788, Pierre Thibert left the mill for eight days to gather his wife and her sister; Mme Thibert soon found work there.[61] Michel Duranton collected his wife and their five children and brought them all to Vidalon-le-Haut in 1786.[62] The husband, it should be noted, was not always the pioneer at the mill. For example, Anne Serenda entered the Montgolfiers' employ three months before her spouse, Blaise Duranton.[63]

Although the chance to pocket a few sous frequently separated worker families, other matters also intervened. In a display of unity with his striking mates, Jean Chatron stomped out of Vidalon-le-Haut in the autumn of 1781. His wife and children stayed behind, and he was able to rejoin them in February 1782 on the strength of a certificat de congé endorsed by Riban of Voiron.[64] Henceforth, in Vidalon's new model shops Chatron was known as "Riban"—presumably a bitter commentary on the means he used to reenter the mill but not bitter enough to prevent him from toiling there for five more years. In May 1787, Chatron *dit* Riban and his wife moved on.[65] Ten weeks later, their four daughters, all of whom were born at Vidalon-le-Haut, employed there, and saddled with the name Riban, quit the mill.[66] One can only wonder if they found work among the men of the "modes."

Once installed at Vidalon-le-Haut, workers were not free of the tugs of family and village. For Vidalon's *patrons*, these connections were a mixed blessing. The carpenter Gagneres brought his wife to the mill, Grenouillet recruited her sister, and the governor Payan's wife evidently encouraged her brother to propose himself for an apprenticeship.[67] Yet, family attachments also pulled workers away from Vidalon-le-Haut. When Claude Artaud learned that his mother was sick, he quit his job and made his way home.[68] Périgord rushed home in 1787 to arrange some family affairs.[69] Workers visited village and kin so frequently, in fact, that a brief furlough to settle family business became the preferred ruse of those hands who intended to disappear without a trace. In November 1784, Jean-Pierre closed the account of the governor Chirol, who had decamped "without saying anything under the pretense of going to arrange a family matter."[70] He never returned. The worker Tissier probably followed the same strategy in 1786. He obtained permission to return home and was not heard from again, leading Jean-Pierre to conclude that his request was "a pretext to flee."[71]

Only obscure clues to the emotional bonds within worker families surfaced in the Montgolfiers' account books. The vatman Joseph Etienne lost two days due to "sickness or the burial of his child."[72] Périgord missed three days while his wife lay in childbirth.[73] Coupy took a payment of 6 livres in

1784 for a service for his dead wife.[74] Two years later, Coupy himself was gravely ill. From his sickbed, he begged the Montgolfiers to retain the 114 livres they owed him; if he perished, he instructed Jean-Pierre to turn the money into a dowry for his stepdaughter.[75] Four days later, Jean-Pierre paid 114 livres to Renan Sage, Coupy's stepson-in-law.[76]

The Montgolfiers treated the worker family as an economic unit. They considered the money earned by its members as family income and usually remitted it all to the father or husband. (Knowing the vatman Joseph Etienne's taste for wine and tobacco, however, Jean-Pierre consigned his wages to his wife "for the nourishment of their child.")[77] In 1784, Jean-Pierre paid Violet for his own labor and that of his two sons; a second account included remuneration of 18 livres to Chirol for his son's efforts.[78] Widows collected their children's pay, such as the 18 livres received by Charbellet for her son Augustin's sweat.[79] Debts, too, were freely transferred within families. Anne Duranton's obligation of 24 livres to Vidalon's *patrons* for furnishings was originally charged against her earnings, but on January 6, 1785, Jean-Pierre noted that "from this day [the debt will be] carried on the account of the wages of her husband."[80] The worker Julien must have tired of looking at his carnet: it reminded him that his wife had borrowed 24 livres from the Montgolfiers to buy grain.[81]

So anxious to bring worker families into the craft and, perhaps, thereby weaken the customary links among journeymen, the Montgolfiers were sensitive to the shifting needs of their hands. They helped Frappa with the expenses of childbirth and granted 3 livres 12 sous to Gagneres for his child's burial.[82] They gave André Baude some work so that he could write to his parents and return to "the bosom of his family."[83] Baude heard from home a week later and left.[84] Not surprisingly, Jean-Pierre considered *his* family's interests as he responded to the workers' problems. He agreed to keep Thibert's wife, a quiet women who knew her craft fairly well, for forty days when her husband quit.[85] Thus Jean-Pierre's pledge to educate young Peroud, the son of a widow burdened with a large family, went beyond his assurance that the boy would learn to read, write, and figure. He also planned to make Peroud "useful to his family [by] inspiring the love of work in him."[86] Finally, one of Vidalon's ordinances claimed that the Montgolfiers had provided their workers with apprentices so they could tend to their own children; in return, these experienced hands were to report any incorrigible greenhorns.[87]

In October 1783, Jean-Pierre responded favorably to an appeal by the father of a recently fired apprentice. The boy got his job back on the condition that his conduct would improve.[88] A year later, the career of another Vidalon apprentice, Rousset, had apparently reached a stormy end. But in

December 1785, when things had presumably settled down, "Rousset . . . whom we had dismissed was presented to us again by his father; and, on the promises of father and son that he would behave better in the future, we rehired him."[89] Despite a few bumps, he was at work in the mill in April 1789.[90] Jean-Pierre made fathers complicit in the submission of their sons; in doing so, he enhanced the stability of Vidalon's bons élèves. Consider the fate of Simon Lombard, a novice from a nearby parish who entered the mill on May 1, 1787. He fled on May 2 but returned a day later, escorted by his father.[91]

Recent scholarship has emphasized the persistence of artisanal production in the era of large-scale machinery and the vitality of small shops among massive factories. Indeed, the forms were—and are—linked in complex production networks. The same holds true for the organization of production and labor discipline. Paternalism was not a stage in the evolution of styles for the governance of workers but a valuable component of multivalent disciplinary regimes. After the Montgolfiers' hiring net snagged choice candidates, their paternalist practices were geared to integrate both novices and veterans into an alternative community of workers. Free of the "modes," this community would have its own norms and reciprocities. The worker family was now to serve as an adjunct of the Montgolfiers' new order and, crucially, no longer as the transmitter of either "modes" or skills. Meanwhile, those mill women and children who had dared to leave the shelter of the Montgolfiers' employ and join their striking husbands and fathers in exile, Augustin maintained in 1781, were "at risk of being one hundred times worse off elsewhere."[92]

Religious instruction, proclaimed one of Vidalon's ordinances, was "the primary duty of fathers and mothers towards their children."[93] On Sundays and holidays, Pierre Montgolfier conducted catechism classes for young workers in the mill.[94] Above all, those hands who had not attained their twenty-fifth birthdays were not to abandon Vidalon-le-Haut without permission on feast days and Sundays.[95] As the "rules to observe in the paper mill of Montgolfier of Annonay" explained, it was "misdirected tenderness" to let children "have their will."[96]

CHAPTER TEN

Wages

"Promise little to your apprentices," Augustin Montgolfier advised his family in 1781. "The less they earn, [the] more they will work."[1] Augustin's counsel reminds us why English entrepreneurs gleefully received news of rising corn prices: "Everyone but an idiot," warned Arthur Young, "knows that the lower classes must be kept poor or they will never be industrious."[2] Young was echoing a widely held conviction: workers would toil until they had obtained their subsistence and then quit, preferring their leisure to ample purses. As Nicolas-Edmé Restif de la Bretonne explained, the "dearness of labor" threatened a populace that "if it can earn what it needs in three days, only works for three days and spends the other four in debauchery."[3] Not every observer, however, shared this opinion. One notable dissenter was Adam Smith: "That a little more plenty than ordinary may render some workmen idle, cannot well be doubted; but that it should have this effect upon the greater part . . . seems not very probable."[4] The physiocrat François Quesnay also rejected the proverbial case for the utility of low wages. He thundered against "the maxims of those cruel men who insist that you must reduce the people to misery to force it to work."[5] Political economists notwithstanding, eighteenth-century English producers did not recruit labor by cutting wages.[6] In 1766, a papermaker in search of a coucher and a layman wrote to Ecrevisse, who was then director of a mill at Rousbreucq: "I will pay them according to their merit, I am not [the sort of] man to hold back their wages."[7]

Still, in an era when producers and philosophers were equivocal about the effects of incentives and high wages, it was daring, if not reckless, to pursue the Montgolfiers' strategy. They provided premiums to the bound men who had replaced the strikers of 1781. There were precedents, however, for the Montgolfiers' approach.[8] In 1688, for example, the master papermakers of the Auvergne conceded that apprentice vatmen, couchers, and laymen would enjoy the same perquisites as the journeymen who performed these tasks.[9] Perhaps the veterans had recognized the threat of low-priced competition and therefore pressed the novices' case. A century later, the Montgolfiers, *without* duress from the men of the "modes," offered annual bonuses to apprentices

already under contract and production incentives to greenhorns who did the work of journeymen. Accustomed to deep pockets, highhanded Augustin might have insisted, the bons élèves would become demanding and debauched. His family, too, betrayed reservations. "They earn, perhaps, too much," said one of them about the vatmen, "and it is that which perhaps renders them less attentive."[10] But Vidalon's *patrons* had to run the risk: they were attempting to create a stable company of tractable workers who pursued their individual and family interests. Lessons in rational calculation, then, did not require free labor—yet another measure of the flexibility of form and adaptability of capitalism at the close of the Old Regime.

The first order of business was to get the bound men to stay. Briquet maintained that apprenticeship in French papermaking lasted an average of three years, a year less than the term specified by the state in 1739.[11] Augustin Montgolfier signed his novices to three-year contracts.[12] The son of the widow Gouton agreed to serve five years as an apprentice, from 1691 to 1696, with a papermaker in the Auvergnat hamlet of Rochetaillée. Young Linossier passed an apprenticeship contract in 1769 with Pierre Montgolfier for five years, but Etienne's bons élèves had to serve six.[13] To receive the reward for the completion of his indentures, a bonus of 300 livres, Jean-Pierre Lorme had to toil at Vidalon-le-Haut for "six full, consecutive years."[14] In 1784, Jean-Pierre Montgolfier paid Chantier 36 livres "without prejudice to the condition that he must still serve me for four years."[15] A year later, Jean-Pierre remitted some funds to Chantier "without prejudice" to the young man's obligation to finish his term.[16]

In effect, Chantier and his counterparts had become bound journeymen. Thus the newcomer who fled before completing his indentures owed Vidalon's *patrons* 60 livres, the basic annual salary (without food or premiums) of a journeyman layman or of a veteran apprentice filling his shoes. Jean-Pierre warned Chantier that he would "repay me one year's wages" if he decamped prematurely and admonished Quinat that "if he leaves early, he will repay me 60 *livres* to compensate me for his apprenticeship."[17] Once Jamet paid this price for his failure to serve "the time promised by his apprenticeship [contract]," he received a certificat de congé.[18] (One can only wonder if the Montgolfiers had the hubris to advertise Jamet as a full-fledged worker.) Surely, the stiff penalty was meant to keep well-trained, skilled men from straying. In addition, the apprentice who quit or had to be sacked was expected to reimburse every sou that Vidalon's *patrons* had advanced to him. Small wonder that many of the tyros who slipped away from the mill left without a word to either parents or *patrons*.[19]

Transforming bound youths into skilled men, the Montgolfiers knew, took

time. Yet, Vidalon's *patrons* could not afford much patience, since many of their greenhorns had to work the vatmen's molds and the couchers' felts. Consequently, the Montgolfiers designed distinctive reward schedules for the bons élèves to encourage the rapid acquisition of skill. There was a certain consistency to most of these schedules, though several reflected individual tailoring. No doubt, the workers had sought these personal premiums and exemptions. But the Montgolfiers gained as well, since the bons élèves were now complicit in the crafting of their own bonds. Moreover, these petitions for individual concessions reminded the young hands that the Montgolfiers held unfettered sway over placement and advancement in the shops. Valençon's dream of a contract that matched Millot's, he quickly discovered, rested on whether his new masters were "content" with him.[20]

Jean-Pierre Montgolfier promised Quinat his meals only "if he behaves well . . . during his apprenticeship."[21] Nourishment was the novice's principal compensation during his first year under contract, which itself was generally the fruit of a successful probationary period. A conventional payment schedule offered the apprentice 18 livres for each of the next five years, plus a substantial bounty of 300 livres when he finished his indentures. Only Michel Chantier, an "excellent worker" according to Jean-Pierre, collected this bonus from April 1784 through December 1789.[22] Chantier's unique status should not be taken as evidence of flightiness; as chapter 13 demonstrates, the Montgolfiers' new men were relatively, even remarkably, stable. Rather, the Montgolfiers' turn to a second payment schedule, a format that mirrored their need to train skilled men quickly, probably accounted for Chantier's lonely achievement. In the second framework, wages increased with skills. Augustin Charbellet's contract called for a wage of 18 livres during his first year as an "*ouvrier*" (likely a layman), 36 livres during his second year, and 60 livres during each of his remaining years of service.[23] Barthélemy Géry labored under the same terms.[24] Accordingly, these hands and Etienne's other bons élèves provided years of relatively cheap labor—labor that was all the more rewarding for the Montgolfiers because it was not enmeshed in the "modes." When maturing hands chafed under these arrangements, however, Jean-Pierre knew his priorities; he adjusted wages, not tenure. Unhappy with indentures that bound him to serve until he was twenty-five years old, Antoine Millot demanded a new contract in 1788. His wages were revised, not his obligation.[25]

In 1739, Versailles mandated four-year terms for both apprentices and journeymen.[26] Eager to promote fresh hands to skilled work, the Montgolfiers' reformed payment schedule conflated the two ranks, provoking the inevitable confusion. Augustin Valençon, who had returned from Thollet's mill after Jean-Pierre's "complaint," was to serve six years as a "*compagnon papetier*"

(journeyman paperworker); if he failed to last, he owed Jean-Pierre 60 livres as compensation for his *"apprentissage."*[27] Jean Texier, a second apprentice who had taken to the road, was rehired at the request of his mother in 1786 "in order to finish his apprenticeship." The day after his return, he was to begin work as a layman.[28] Evidently wary of Vidalon's entanglements, one young hand agreed to put in a year without wages when he filled a skilled station instead of passing an apprenticeship contract.[29] The bons élèves, then, had entered an alternate hierarchy where the Montgolfiers—rather than the men of the paperworkers' association—welcomed them into the tasks and company of journeymen.

It was the Montgolfiers' reworking of conventional classification that explains why the apprentice Jean Frappa enjoyed an annual bonus of 24 livres in 1785.[30] Neither fish nor fowl, Frappa finished his apprenticeship while performing the work of a skilled journeyman. His rising premium was geared to teach him that long seasons of regular hours and output, however taxing, had their rewards. "Instead of paying the workers as in the past, when each vatman, coucher, and layman had the same wage," wrote one of Vidalon's *patrons*, "I feel that it would be advisable if wages would be in proportion to the talent of each man." He realized that the workers he assigned to the vat crews that fashioned expensive papers would earn more. Rather than include this "surplus" in the regular wage, he planned to retain it, create an annual bonus, and thereby enhance persistence. Should two workers possess equal talents, the man who had put in the most years at the mill would receive the more lucrative post.[31] As slackers fell back and skillful men rose, Vidalon's *patrons* would possess a visible spur to proper work and conduct, and a subtle reminder of who alone judged it all.

Whether the Montgolfiers' implemented this scheme remains uncertain. They did offer an annual bonus of 36 livres to one journeyman in February 1786; he had to sweat in the mill for another year and continue to please his bosses to obtain it.[32] Nine months later, Jean-Pierre promised a premium to Jean Peschi as long as his good conduct persisted.[33] Jean-Pierre also threatened the besotted Chatron dit Riban (who seemingly had reason to drink) that his annual supplement would vanish "from this moment [in February 1785] if he drinks in the workshops or makes a din in the household."[34]

The Montgolfiers' construction of a new building for the beaters and the vats they fed complicated their proposed ranking of production units. So did their reliance on carefully selected journeymen as well as home-grown hands as the 1780s progressed. In effect, they had two sets of tools and men. Typically, they resolved all this by blending "licit" customs of the craft with incentives and classifications of their own making. Furthermore, their plans and

payment schedules revealed their belief that the meaningful sphere of skill—the conceptual element—belonged to them.

In December 1788, Jean-Pierre Frappa learned the terms of his new wage schedule. He had just completed his apprenticeship and was capable of handling any skilled post. When he toiled around the vats that drew pulp from the stamping hammers, his annual wage as a vatman was 90 livres, as a coucher 78 livres, and as a layman 60 livres. When he switched to pulp furnished by the beaters, his yearly wage was 60 livres, regardless of his task.[35] The worker Quay left Vidalon-le-Haut in 1788 with 51 livres 16 sous in his pocket. He had spent four and a half months at vats supplied by mallets at 7 livres 10 sous per month and a little more than three and a half months around production units fed by the beaters at 5 livres per month.[36] Whether the worker was a tramping veteran or a bon élève, these schedules held. At the vats linked to stampers, the itinerant exercised his own skills. In the Montgolfiers' new shops, however, journeyman and novice alike practiced skills taught, at least in part, by the *patrons* and earlier, Ecrevisse. Here the veteran and the tyro were both bons élèves and, hence, earned at the same rate; even more, the Montgolfiers had dispensed with distinctions in pay for the vat crew precisely because *all the skills* the workers put into play were determined by the *patrons*—that is, belonged to the Montgolfiers. At the beater-vats, greenhorns and graybeards were simply interchangeable parts, the manual expressions of the Montgolfiers' mastery. There was no room for the "modes" or skills transferred from father to son in the Montgolfiers' new shops. Instead of deskilling, the Montgolfiers were discounting the value of experience, and defining, transmitting, and rewarding skill on their own terms.

The promise of annual premiums to journeymen like DuBesset disappeared when they sweated around Vidalon's beater-vats. When DuBesset arrived at the mill in 1787, Jean-Pierre offered him an annual wage of 90 livres plus a bonus of 30 livres if he lasted the year at a hammer-vat, but only 60 livres without the prospect of a bounty when he labored in the new shops.[37] Still, the Montgolfiers enjoyed less latitude when tramping hands toiled in conventional ways with conventional pulp. Evidently, Vidalon's *patrons* concluded that itinerants, however docile and hand-picked, would not forgo the perquisite known as the *"avantages communs."* So they transmuted it into an annual bonus and stretched out whatever good will—and behavior—it prompted.[38] Vidalon's *patrons* counted on customary workloads, payment by results, and a host of incentives to teach their hands that steady work and reliable conduct paid. Thus the heart of their wage system, as for English and French papermakers in general, was the "day" or "day's work"—daily quotas.[39]

The Montgolfiers sought the seamless production of quality paper. They shared the outlook of the pioneer English industrialists who concerned themselves more with the maintenance of regular output than with spectacular, short-term yields.[40] While they fretted about reams and hours lost en débauche, they also feared hasty, substandard work. Etienne Montgolfier even recoiled from one incentive because of its very success. In their eagerness to pocket a few extra sous, he recalled, the workers sacrificed both their health and the quality of their wares.[41] With their discriminating customers and elevated prices, the Montgolfiers were in no position to surrender quality for quantity.

Perhaps the best way to approach the "day" at Vidalon-le-Haut is to explore a week's worth of work in depth. Fortunately, one leaf survives from an early effort by the Montgolfiers to track output apart from their other accounts. It depicts production at the "old vat" during the week of September 10, 1781. Whereas workers in early modern Europe commonly abstained from their labors to venerate "Saint Monday," Monday marked the peak of production at the old vat. The team made twenty-eight posts of paper (about 3,500 sheets), eight posts above their quota. For this prodigious output, the vatman earned 1 livre 7 sous 6 deniers; the coucher, 1 livre 5 sous 1 denier; and the layman, 1 livre 2 sous 6 deniers. On Tuesday, the team fashioned only eight posts. A broken waterwheel, a punctured vat, or a dry race, among other problems, could have disrupted Tuesday's papermaking. Whatever the reason, the Montgolfiers termed Tuesday a "breach day," thereby taking responsibility for the shortfall. They released the vat crew from its routine duties and gave the men reduced wages, so long as they spent the remainder of the day on other chores in the mill. On Wednesday, Friday, and Saturday, the team met its quota, thereby netting daily earnings of 1 livre 1 sou 6 deniers for the vatman; 19 sous 1 denier for the coucher; and 18 sous 6 deniers for the layman. On Thursday, the men at the old vat missed their quota by a single post. For the failure, the vatman lost 3 sous 6 deniers; the coucher, 2 sous 4 deniers; and the layman, 6 deniers. Each post in a routine day added 1 sou 1 denier to the vatman's purse; each post above this total brought him 10 deniers; and each post under his quota cost him 3 sous 6 deniers. The stick outweighed the carrot in proportion to the Montgolfiers' interests: they encouraged steady production rather than extraordinary efforts.[42]

To ensure uniform output when they were shorthanded, the Montgolfiers and their neighbors, the Johannots, turned to a customary arrangement known as the *"porse à deux."* These posts were the work of a vatman and a coucher without the aid of a layman. Since these men had to separate the sheets and felts and lug the newly minted paper to the press, Annonay's *patrons* shaved

four to eight posts from their quotas.[43] The workers' purses did not suffer, they were not driven to produce slipshod reams, and their attention to the "day's work" was not interrupted.

"Avantage," as Desmarest defined it, was the "exceptional work of the vat crew" that generated "a certain augmentation" of its wages.[44] The workers who sweated for this bounty called the extra posts *"bâtards."* Seasonal opportunities, particularly abundant water, often led manufacturers to prolong the journeymen's work. In their haste, the papermakers resorted to linen that had not endured sufficient fermentation or pulping. Since this material yielded inferior paper, the state, in a late expression of Colbertian zeal, barred the fabricants in 1739 from imposing the "extraordinary tasks called *avantages*."[45]

Within Vidalon's "day," a piece-rate system inscribed in a fixed workday, the Montgolfiers had a second use for the term *avantage*. They proposed wage calculations on the basis of "*avantages* for a third of the [regular] workday."[46] In about fifty-five days of work at Vidalon-le-Haut in 1786, Alexandre Doire fashioned 224 *"porses d'avantage"* (bonus posts of paper). As he was expected to make twenty to twenty-six posts of paper daily, Jean-Pierre had, in fact, treated one fifth of Doire's output as "exceptional" work.[47] Certainly, the Montgolfiers offered the conventional bonus of their trade for overtime work; evidently, they also borrowed the term *avantage* to stimulate quota-making and quality production. Doire and his companions only obtained this premium if they did not turn out too many "broken sheets."[48] When Louis Vallet left the Montgolfiers after four and one third months of work in 1787, Jean-Pierre paid him 32 livres 10 sous in fixed wages (at 7 livres 10 sous per month), 10 livres 6 sous 3 deniers for 275 porses d'avantage, and 6 livres as compensation for his avantages communs (the bounty Vidalon's *patrons* had reconfigured as the reward for sustained service).[49] Vallet was a favored worker: his twin premiums suggest that he both reached and exceeded his quotas.

This terminological thicket makes it difficult to compare the earnings of Vidalon's hands with other paperworkers, much less journeymen in other trades. For a year's work of three hundred days, according to one of the Montgolfiers' many calculations, the vatman would draw a base salary of 90 livres (6 sous per day); the coucher, 66 livres (4 sous 5 deniers per day); and the layman, 60 livres (4 sous per day). Each journeyman would also receive food valued (by the Montgolfiers) at 198 livres (13 sous 7 deniers per day).[50] In a second scheme, which included a daily incentive of 6 sous, Vidalon's *patrons* figured that a vatman could make 382 livres 5 sous per annum.[51] Finally, the Montgolfiers' audit of their expenses at the terrace vat in July 1780 revealed that the vatman earned about 30 livres in food and wages for twenty-five work-

days.[52] For comparison, the journeymen paperworkers of the four mills in the Languedocian town of Mazamet claimed in 1788 that they earned either 7 or 9 livres per month in wages, plus their food.[53] But what of their bonuses? François Millot received 542 livres 12 sous 4 deniers "over and above [his] basic pay" for six years of work at Vidalon-le-Haut. (Jean-Pierre deducted 16 livres more for sick days and débauche.)[54] In a related craft, Etienne Montgolfier's friend Jean-Baptiste Réveillon, the producer of elegant wallpaper in the faubourg Saint-Antoine of Paris, paid his skilled hands 30–50 sous per day.[55]

According to Pierre Léon, the laboring poor of the Old Regime earned up to 20 sous per day, the labor aristocracy more than 30 sous, and a large group of skilled and semiskilled men (probably including most journeymen paperworkers) fell in between.[56] But the realities of production in eighteenth-century France, of seasonal disruptions and workless days, limit the utility of such formulations. Occasional grants further complicate comparisons, such as the "indemnity" of 12 livres that Jean-Pierre provided Filhat for the transport of his "furnishings" to the mill.[57] Annual bonuses, customary gifts, and a few sous for such routine chores as refurbishing the vats all edged Vidalon's veteran apprentices and journeymen closer to the upper strata of worker pay. But what distinguished the mill even more was its long papermaking season—and, perhaps, the varied diet enjoyed (?) by its male hands.

Papermakers eager for skilled hands frequently "debauched" (enticed) workers away from their confrères. They poached journeymen from their proper masters with the promise of high wages, a practice Versailles had declared illegal in 1739. The brazen manufacturers of Montargis, near Paris, even sent recruiting agents into the Auvergne, which refreshed the animosity between these industry centers.[58] At Mathieu Johannot's mill, the vatman's annual base pay was 90 livres; the coucher's, 66 livres; and the layman's, 60 livres—amounts identical to those paid by the neighboring Montgolfiers.[59] The decree of 1739, after all, contained no penalties for producers conspiring to hold down wages. Presumably, Annonay's papermakers had acted in concert to prevent travel up and down the Deûme: Vidalon's *patrons* even began one of their ordinances by noting the pay for sorting superfine rags "chez M. Johannot."[60]

To keep valued hands in the mill, the Montgolfiers offered pay increases. The governor Payan, whose wage was 110 livres during his third year at Vidalon-le-Haut, was entitled to 120 livres for his fourth.[61] Jean Texier, the apprentice layman rehired on his mother's pleas, had wages of 18 livres for his first year, 30 livres for his second, 42 livres for his third, and 54 livres for his fourth.[62] Such arrangements served much like advances, rewarding and setting the terms for work that was yet to be accomplished. The carpenter

Antoine Gagneres would receive a wage of 300 livres per year, lodging, and candles; in exchange, he was to labor from 4:00 in the morning until 7:15 in the evening.[63] The governor of the Hollander beaters obtained an annual wage of 120 livres, a bonus of 12 livres, and piece payments for work carried out on Sundays and holidays. In return, he had to supervise three machines during a twelve-hour shift while performing various other tasks.[64] When the Montgolfiers hired several ouvrières papetières in December 1781, they explained that the women must toil from 4:00 A.M. to 7:00 P.M.; in exchange, their new hands would obtain 5 sous per day for food, soup to soak their bread, and wages to be arranged when the Montgolfiers knew their *"savoir-faire."*[65] A month and a half later, Vidalon's *patrons* offered an annual wage of 30 livres to both Faurier and Margerier, but Faurier could expect a raise of 6 livres after a few months of service. For a year's work, Mayol would receive 33 livres.[66] Anne Crêpe really caught the Montgolfiers' fancy. At first, they granted her an annual wage of 24 livres, then promised a boost of 6 livres in April 1782, and finally matched Mayol's pay in July.[67] Within a fairly narrow range, each woman had a personal pay scale, echoing Augustin's claim that "each worker [in Rives] has a different wage [*prix*]."[68] Self-interest would prevail over solidarity, the Montgolfiers imagined.

The Montgolfiers built their wage system on the traditional footings of the craft: the "day," premiums, and apprenticeship. Yet, they ceaselessly tinkered with these forms and practices, adapting them to both familiar and novel purposes. Their embrace of wage incentives seems more pragmatic than philosophical. They were tutoring *"ouvriers-élèves,"* in Etienne's succinct phrase, in the rules of the market game, their game. They adjusted custom and its terminology to speed their novices' journey into the ranks of the skilled. Accordingly, the process of proletarianization at Vidalon-le-Haut *endowed* certain workers with skills—skills determined by the Montgolfiers, as were the rewards for mastery and performance.

Like England's early factory hands, Vidalon's new men "had to be made obedient to the cash stimulus, and obedient in such a way as to react precisely to the stimuli provided."[69] When Jean Bacon claimed that his wages were insufficient and moved on, Jean-Pierre characterized him as a worthy lad "who knows nothing yet."[70] In November 1786, Jean Roderic left the mill without a word. He had insisted on higher wages, but Jean-Pierre turned a deaf ear to his demands. Roderic was a "decent enough lad" but dull, a failing that stemmed from his roots: Jean-Pierre portrayed him as an "Auvergnat in all the force of the term."[71] Bacon and Roderic had obeyed the cash stimulus much too well for Jean-Pierre Montgolfier, who preferred to attribute their departures to inexperience and stupidity.

CHAPTER ELEVEN

Discipline

"An ingenious artist (Monsieur Montgolfier) contrived three figures in wood to do the work of the vatman, the coucher, and the layer; but, after persevering for six months, and incurring considerable expense, he was at length compelled to abandon his scheme."[1] Like the wooden workers depicted in the *Encyclopédie*, Monsieur (probably Joseph) Montgolfier's automata, if they existed, represented an idealized, lifeless world of work.[2] Drained of the journeymen's disputes, humor, and rites of passage, papermaking might become clockwork or, at least, submit to the discipline of the clock.

Unlike M. Montgolfier's short-lived experiment with automata, Vidalon's *patrons* persisted in their effort to create new model workers. Yet, precise regulations for their new hands' tasks and restlessly revised output schedules did not qualify the Montgolfiers as proto-Taylorists. Above all, Taylorism turned on an ever-widening division of labor, in which the skilled man and his judgments would be rendered obsolete. Subordinating such men to disciplinary designs, no matter how relentless, was not Taylorism.[3] In fact, the Montgolfiers were training swarms of hands in full, rich skills—without the "modes," of course. Vidalon's *patrons* sought to separate this skill from its culture, thereby deregulating their art and its improvement. Their "scientific management" aimed at a more rational technology, a series of experimentally verified tasks carried out by skilled men. Their new order would overturn the rule of thumb as well as the tyranny of the "modes." Meanwhile, the Montgolfiers themselves would ensure that all gears meshed: their workers would labor quietly and mechanically, as if they were made of wood.

Like Franco Venturi's eighteenth-century cosmopolitans, the Montgolfiers intended to serve as beacons of advanced technology and labor relations while "desiring to transform and enlighten [their] little corner of the world."[4] Vidalon's recognition as a manufacture royale merely enhanced this conviction. In 1786, the Montgolfiers responded to a request from an Ambertois producer for "enlightenment" on a number of matters. Their answer considered the wiring of molds, the number of quires and posts in a routine workday of

cartier fin, and the steadfast resistance to the "modes" that lured flocks of townsmen and countryfolk to Vidalon-le-Haut for apprenticeships.[5]

Even the Montgolfiers' impersonal standards for work and conduct, the hallmark of "the new factory discipline of the eighteenth century," bore the imprint of old conflicts within the craft. Four years after the lockout, a new ordinance reminded Vidalon's hands that they would "introduce nobody into the household or the workshops without the permission of the master."[6] Pioneer industrialists everywhere labored mightily to control mill gates and shop doors. Oberkampf did so to monitor punctuality and discourage theft at his calico-printing works.[7] As a lesson in the thrifty use of time, Oldknow and the Braids in England locked the factory gates and the workrooms, thereby excluding those who were minutes late.[8] The Montgolfiers also demanded prompt attendance and a lengthy workday, but it was the rentes that led them to press each governor and sizerman to bar unauthorized visitors from his shop.[9] Evidently, the workers countered with the terms of the subsidy for the beaters. Though their Dutch shops were to remain open to any manufacturer, Vidalon's *patrons* cautioned their bons élèves to bring no one into the mill without permission, "even under the pretext of displaying [their] skills."[10]

The assignment of work at Vidalon-le-Haut, said the Montgolfiers, was their business, and their business alone. Whereas the "modes" affirmed the journeyman's right to select his post, Vidalon's *patrons* proclaimed that it was the master's "need and his advantage which must determine the places that the *ouvriers* and *ouvrières* occupy, and they should comply."[11] Managing the flow of work and workers at eight vats, double the number of a few years earlier, was complicated enough. In January 1786, the worker Tissier gave his notice. He wished to choose his own station and to avoid the company of Grenier, a second Vidalon hand. He was absent from the mill during the morning and afternoon of January 30 and perhaps spent more time en débauche. He received a payment of 12 livres on February 8 and, a week later, withdrew his notice. He also pledged to abide by the règles de la maison, which included the provision that Vidalon's hands must toil at the jobs and with the men specified by the *patrons*.[12] Jean-Pierre allowed the humbled Tissier to remain.

The worker Poynard, the bon élève who had resisted the journeymen's imposition of an appricticeship fee in the autumn of 1781, was a man of strong will and drives. On May 24, 1784, Jean-Pierre sacked him. He had attempted to rape a girl, and the criminal complaint against him had found its way to the mill. With the aid of 9 livres provided by the Montgolfiers, the matter was "settled." Poynard then pledged "to behave himself better in the future," and

Jean-Pierre reinstated him on May 26. Four months later, Poynard failed to appear at his vat. Offended by a change in his assignment, he did not inform the Montgolfiers of his absence. When Jean-Pierre put another man in his place, Poynard demanded his congé. Jean-Pierre lost no time in ridding the mill of Poynard, who moved on with an honorable discharge. It was one thing to assault a girl, another to challenge the Montgolfiers' authority over their shops.[13]

To match the orderly, rationalized workrooms of the *Encyclopédie*, the Montgolfiers had to banish the hurly-burly of everyday working life from their mill. They linked lewd language with tumult, and dreamed of putting an end to both: "The *ouvriers* and *ouvrières* will live in peace together, neither quarreling nor swearing at all nor saying obscene words; and if anyone has contracted the habit [of vulgar speech] he will try to rid himself of it; and if he persisted in this bad habit we might be obliged to dismiss him."[14] Yet, the Montgolfiers did not fire any man during the 1780s for obscene speech. In any disciplinary code, such provisions constitute a wish list rather than a reflection of actual practice. They did, however, figure in a larger vision: if swearing, the "modes," and other distractions were expelled from the mill, work at Vidalon-le-Haut might assume clocklike precision.

As early as 1688, the state set standards for the daily output of the kingdom's papermaking vats.[15] In practice, however, the journeymen lifted the molds and couched the freshly minted sheets until the *ouvrage* (pulp) ran out.[16] In place of this changeable pattern, Vidalon's *patrons* established a regular *workday*. Although its inner divisions were clear cut, they shifted, if only slightly, from season to season and from one mill ordinance to another, as did quitting time. The governor of the Hollander beaters sounded the bell at 3:45 A.M., signaling the beginning of work. Fifteen minutes later, the vatman dipped his mold into the warm, watery pulp, an action he would repeat as many as four thousand times that day. Breakfast began at 6:00 (or 6:15), lunch at 11:00 (or 11:30), the collation at 4:00, and supper, followed "immediately" by the evening prayer, was served at 7:00 (or 7:30). Each of the principal meals lasted for forty-five minutes; the snack took thirty.[17] Consequently, Etienne Montgolfier based one wage calculation on an "effective workday of 13 hours."[18] By breakfast, declared one Montgolfier mandate, the vat crew must fashion five posts of paper.[19] A second code announced that Vidalon's hands should take no more than forty-five minutes to complete each post, itself a variable quantity of sheets determined by the format of the paper.[20] Still, Vidalon's regimen gave the workers some sway over their personal time. Although one mill ordinance maintained that the workers should retire at

8:00 P.M., Jean-Pierre informed a new hand in 1786 that he should turn in at a "decent" hour.[21]

French papermakers fumed about the journeymen's preference for late night and early morning work. In Ambert and Thiers, the paperworkers began their labor at midnight or 1:00 A.M. and usually put down their tools at noon or 1:00 P.M.[22] In 1794, the workers at the big Johannot mill petitioned for more candles because "each worker [made] a third or more" of his wares without the benefit of sunlight.[23] Contemporaries and historians have offered several explanations for the paperworkers' unusual hours: a desire to evade the worst heat of the day; to spare some daylight hours for field work; to preserve afternoons for the tavern and confraternity; or to seek the cover of darkness for the theft of rags and reams.[24] And the description of the boisterous journeymen paperworkers of Chamalières as they made their midnight journey to the mills, doubtless spoiling for a fight, calls to mind yet another interpretation: what better way for the men to demonstrate that they did possess a distinctive essence?

Whatever the case, the bosses blamed night work for shoddy wares and larceny, so the state stipulated that the workers must perform half of their labors after noon.[25] Like much of the government's regulation of the trade, this mandate was hardly worth the paper it was written on. In 1772, certain paperworkers in Thiers went on strike to protest their masters' effort to economize on candles and oil by locking their mill doors until 4:00 A.M. At first, the journeymen of the town were split, but the hesitant soon leagued with the strikers. The fabricants retaliated by engaging scabs calculated to wound the journeymen's pride: women, children, and veterans from other provinces. Bullied by the strikers, the experienced "foreigners" were the first to disappear. The manufacturers then sacked the women and pressured the *cabaretiers* to cut off the strikers' credit. The subdélégué intervened, but to no avail. Finally, the producers caved in: they agreed to humble themselves and invite the strikers back, to compensate the journeymen fully for the two months of work they had lost, and to open their mills at 3:00 A.M.[26] A regular workday that began shortly before dawn was no mean achievement in Old Regime papermaking.

"With the agreement of the master," Vidalon's *patrons* permitted "very strong workers" who wished to earn something extra to begin their workdays at 3:00 A.M. Nevertheless, these men were expected to knock off "at the same hour as the others." When scarce water idled one vat, the Montgolfiers doubled up teams at another and worked them in two shifts punctuated by noon and midnight. They also modified the meal schedule, so that the journeymen

"will not be deprived of food for such a long time."[27] Yet, Vidalon's *patrons* betrayed no anxiety about restoring their daylight schedule; after all, their workers themselves observed that "we . . . will finish our day at 7:30."[28]

The Montgolfiers were especially clear about the source of the light in their shops. Daylight alone would suffice from the first of May through the first of August. For the remaining months of the year, Vidalon's *patrons* drew up the following schedule for work by candlelight: August, 0.5 hour; September, 1–2.25 hours; October, 4.25 hours; November, 5.75 hours; December, 6.75 hours; January, 5.75 hours; February, 4.75 hours; March, 3 hours; and April, 2 hours.[29] Accordingly, Vidalon's production schedules remained in effect year-round, an uncommon pattern when cycles of manufacture, not to mention construction and farming, followed the seasons. Even the shops of the punctilious Montgolfiers, however, were imperfectly isolated from the harvest calendar: the besotted Chatron dit Riban forfeited two days at the mill for the grape harvest in 1783, and two hands who left to pick grapes during a dry spell in 1786 pledged to "return with the rain."[30]

Vidalon's *patrons* even established routines—and penalties—for those tasks unaffected by daily quotas. The *releveuses*, the women who prepared posts for the press, had to clean their benches daily and sweep their shops twice weekly.[31] The women of the finishing room were to count and sort the sheets separately.[32] The carpenter Clemenson agreed to put in four extra days at the mill as compensation for several short days.[33] Such precision fit a family enterprise whose patriarch wrote that it was a twenty-minute walk from Davézieux to Annonay.

In 1778, three years before the lockout, the Montgolfiers penalized five workers for a "half-day lost, having been drinking."[34] Seven years later, Jean Texier gave up a few sous for "a quarter-day lost in the month of June; went to town."[35] The Montgolfiers' attention to their workers' hours and output did not start with the lockout and the expulsion of the "modes"; they had been governing a synchronized production process in a large fabrique for many years. They had long known that time was money, as did their former workers who struck amid vatfuls of pulp.

After 1781, the Montgolfiers' task was to instruct green hands from farms and cottages in what avantages, bâtards, and customary "days" had taught men reared in the craft. Thus Payan, a governor, regained his job at the mill in 1787 only after he promised to labor sixteen hours a day and never leave the mill during this time without permission.[36] Such education, of course, offered fertile ground for turnabout: Vidalon's hands grumbled about workloads that were "a little too fatiguing."[37] They even got production schedules revised in their favor. A few days after the Montgolfiers issued a piece-rate formula

in October 1789, for example, they altered both the goals and the payments for certain formats to the workers' satisfaction.[38] Evidently, both Vidalon's *patrons* and workers treated new quotas as trial balloons: so long as making paper rested on the know-how and shoulders of skilled men, the bosses' goals effected little without the workers' consent. When Vidalon's workers spoke of quotas that had been "arranged and agreed," they meant it.[39]

E. P. Thompson challenged historians to consider not only external compulsions to use time profitably but also "the inward apprehension of time" among working people in the past.[40] He rightly emphasized the gap between the bosses' intentions and programs and the workers' own notation of time and assessment of its meaning and worth. Surely, however, there is a remarkably specific sense of the value of time, at least in terms of piles of paper, in the petition entitled "representation to [Vidalon's] master of the qualities of paper that we will fabricate at three quarters of an hour per post."[41] It is also tantalizing to note that Pacquet, a governor who passed up Sunday and holiday reveling in favor of paid work, spent 48 livres in 1788 to buy a watch.[42]

The Montgolfiers numbered reams as well as workers. In both cases, Vidalon's *patrons* sought to simplify audits and fix responsibility. For instance, each paper-sorter had a number, and the steward of the finishing room was supposed to inscribe it on every ream she handled.[43] All of Vidalon's tools and fixtures bore numbers. This accounting became so refined that the implements associated with each vat carried its number, and the instruments destined for use by the *ouvrier*, *coucheur*, *leveur*, and *apprenti* bore the letters *o*, *c*, *l*, and *a*.[44] One mill ordinance even required the governor of the stamping mallets to have a pair of sabots for his work in the drying room and a clean pair for his toil around the pourrissoir.[45] Otherwise, he risked blame for filthy pulp or soiled sheets.

Establishing standards for the care of Vidalon's tools and furnishings consumed many sheets of Montgolfier paper. Often, pioneer manufacturers had to replace an artisan's pride in the tools of his trade with regulations concerning the maintenance of the mill's equipment. Since papermakers had always supplied the molds, felts, and rags, Vidalon's *patrons* simply wanted apprentices and journeymen alike to treat their property carefully.[46] The women of the finishing room, said the "plan to establish order in a paper mill," "will answer for the tools that have been entrusted to them."[47] Inattention would be punished: in 1789, the sizerman Mironnier learned that he would be liable for the damage caused by his truancy.[48]

Even the producers of cheap wrapping and speckled writing paper knew that soiled sheets reduced the value of their wares. Suppliers of discerning

customers, Vidalon's *patrons* were obsessed with the cleanliness of their hands. The women of the sizing room will have "clean hands"; the apprentices of the finishing room "will keep themselves clean, [their] hands always washed"; and the ouvrière who soiled sheets was liable for her "slovenliness."[49] The mill had several toilets—evidently a fresh experience for many hands. Vidalon's *patrons* felt obliged to command "the *ouvriers* and *ouvrières* [to] make their excrement in the spots that we have pointed out to them."[50] Remarkably, the Montgolfiers observed that "cleanliness being useful to health, the workmen will pay attention never to deposit any rubbish in the workshops nor on the terrace which is in front of their room." The violator of this rule would fork over 10 sous to the Montgolfiers, who would distribute the funds to the apprentices charged with sanitation.[51] (Doubtless, this unpleasant work served as one test to determine if newcomers deserved apprenticeship contracts, a trial not unlike that of the Slovakian apprentice paperworkers who performed maid's chores.)[52] Ironically, workers eager to bathe also worried Vidalon's *patrons*, especially if they took the water of the post-washers before they finished their tasks.[53]

Broken tools and breaches in their authority came together in the Montgolfiers' warning that "the *ouvriers* and *ouvrières* will never go into one or another of the shops where they have no business on pain of being responsible for the damage, disorder, or dirtiness."[54] Certainly, petty theft plagued the Montgolfiers. Every Saturday night, the vatmen had to turn in their valuable molds.[55] The female hands who had borrowed the masters' furniture must return it. Without "express permission from the master," the workers will refrain from employing paper, "pulp, string, ropes, cloth, wood, and other objects" for their "private use."[56] As soon as he completed "the master's work," the journeyman or apprentice must return his tools to their proper place; a fine of 1 sou awaited any hand who failed to do so.[57] Habitual larceny shortened several careers at the mill. For example, Jean-Pierre fired Barthélemy Mayol in 1787 for wastefulness and "several small thefts."[58]

Despite the workers' complaints about its quality, much of this pilferage involved food. One mill code demanded that an ouvrière be stationed in Vidalon's kitchen at all times; furthermore, she was not to abandon her post under any pretext until another woman arrived.[59] Food disappeared nonetheless. In September 1784, Jean-Pierre fired Claude Pigeon, a worker who had stolen some food.[60] Jean-Pierre never trusted François Charbellet, whom he rehired in August 1784 to work beside his mother. One false step, Jean-Pierre threatened, and Charbellet would be shown the door.[61] In October 1785, he was caught with a sack of the Montgolfiers' chestnuts, fined 1 livre 10 sous, and cautioned that the next episode of "scamp-like behavior [*galopinerie*]" would

find him "irrevocably" out of the Montgolfiers' employ.[62] Chanteloube was less fortunate, perhaps because of Pierre Montgolfier's stern religiosity. He lost his job when he ate grapes and did 1 livre 16 sous' worth of damage in the Montgolfiers' vineyard during Mass.[63]

One woman was sacked when Jean-Pierre caught her stealing rags from her co-workers.[64] Whether she did so to boost her earnings or to sell the linen is unclear, but there is little doubt that the theft of rags inflamed Jean-Pierre. Old linen was never cheap, and the collection of this material by traveling tradesmen ensured a ready market for pilfered cloth. When he caught Marie Maron, a second Vidalon hand, on her way to deliver a bundle of stolen rags to a "fence," Jean-Pierre was furious. He maligned Maron as "a totally corrupt subject" and charged that "even to be in [her] company [was] dangerous." An inspection of Maron's belongings turned up more linen and some other articles taken from both the Montgolfiers and her fellow workers. Jean-Pierre dismissed her but retained her pay and some possessions to force Maron to identify her accomplice. Should she fail to provide a name, Jean-Pierre planned to turn over her earnings and property to charity.[65] A thief with confederates always received a tougher penalty than a lone culprit. For instance, Augustin Bonnetton lost 12 livres for burning the Montgolfiers' wood, but a second hand lost his job when he was spotted among several men making a fire with the Montgolfiers' lumber.[66] Even when the "modes" played no role in illegal acts, Vidalon's *patrons* were particularly provoked when these incidents included several workers or assistance beyond the mill.

Workers occasionally sabotaged Vidalon's tools and wares, but these were individual actions, more a matter of personal pique than concerted Luddism. Racine, an apprentice, earned a fine of 2 sous in 1800 because he heaved an implement into a vat warmer; in the same year, another hand surrendered 3 sous for throwing a wad of paper into a boiler.[67] Perhaps some of the damage Jean-Pierre chalked up to indifference was also intentional: when Bombru spoiled some merchandise, Jean-Pierre preferred to attribute the loss to negligence rather than "ill-will."[68]

In their campaign against larceny, waste, and sabotage, the Montgolfiers relied on a host of informants. Their customers' complaints about deteriorated paper, incorrect counts, and poor sorting were one source of information about the reliability of Vidalon's hands.[69] There was something less than Foucauldian surveillance in the pleas by Marianne Montgolfier's brothers that she take her information-gathering duties seriously; she was to alert them if she saw anything amiss in the mill.[70] And Bon could expect an extra sou per day as long as he kept an eye on the work of Bonnetton, a fellow governor.[71]

Certainly, the Montgolfiers did not subcontract their authority. They had

struggled too long with the "modes" and the combination that sheltered these customs to deed away their sway. They did, however, entrust the workers performing certain tasks with limited responsibility for their mates and their work and, perhaps, thereby practiced a bit of divide and conquer as well. The vatman should take pains to see that the other members of his crew did their work properly.[72] "It is agreed," said Vidalon's *patrons*, "to give the vatman a *sort* [my italics] of authority over his entire crew which he will organize; and we will complain to him when anything is wrong."[73] Without the power of enforcement, such authority amounted to little more than the requirement to consult or to complain to the real boss—and failure to do so could be costly. When he received "badly pulped" rags, the vatman was to notify the Montgolfiers or pay the price.[74] Next in line, the coucher was to alert the *patrons* when his posts of paper were unclean; furthermore, "he will take care to count the post [as] guarantor that it is complete."[75] Even the lowly layman could be called to account for an apprentice with dirty hands or clothing caked with coal dust. At the other end of the production process, the sizerman who supervised the paper-sorters was responsible for their errors.[76] Vidalon's *patrons* even expected longtime hands to impart *"bons principes"* to novices. Evidently, however, the veterans often turned the greenhorns into pot boys.[77] Thus the Montgolfiers encouraged their apprentices to obey the vat crew at work but threatened to boot out any newcomer who followed a command to fetch wine.[78] They made good on this threat in January 1785, when they dismissed Jeannot Pigeon: he had fueled the workers' revelry by "hiding" their wine.[79]

None of the Montgolfiers' workers could complain about the strap or even an occasional cuff to the ear. The mill women especially had more to fear from their fellow hands. Two weeks before the Montgolfiers fired Chanteloube for his sins in their vineyard, they had expelled him for throwing a rock at Le Bon's wife that "gravely" wounded her. Three days later, after Le Bon made an appeal for "mercy," Jean-Pierre took Chanteloube back.[80] (One can almost see the widow Chanteloube, an old Vidalon hand and the boy's mother, looming behind this plea for forgiveness.) "Headstrong" DuMoulin "respond[ed] badly" when Jean-Pierre "showed him his duty." "Always running after the girls," he "went fishing" in the wrong place. His catch was a gunshot, which convinced Jean-Pierre to forgo rehiring him.[81]

Violence was so common among the workers that Jean-Pierre dismissed Fagot for injuring another hand by pelting him with stones yet did not shut the door on his return.[82] And Duranton's departure in May 1785 rested on his refusal to work without apprentices; they had all fled his station because he "mistreated" them.[83] (Since these boys were used to rough handling, one can

only imagine what Duranton did.) Tumult clearly persisted at Vidalon-le-Haut, despite the Montgolfiers' new regime. Still, all but a pair of these incidents were the work of individuals, and none appears to have involved the "modes."

On November 28, 1785, five Vidalon hands—Jean Texier of Serrières, Jean Chantier, Chaumier, L'Horme, and Jamet—decamped to Annonay, doubtless in search of wine. Three of the young men soon ended their lark, but Jamet demanded and received an honorable certificat de congé. L'Horme refused to pick up his tools, fled without a written discharge, and failed to reimburse Vidalon's *patrons* for his apprenticeship. Meanwhile, on November 29, Jean Texier of Annonay took off in the company of François Charbellet and three unnamed apprentices. The anonymous trio quickly got back to work, but Jean-Pierre ran off Charbellet and Texier of Annonay; Charbellet's "insolent remarks" had particularly offended him. Yet, he concluded that Jamet, though "a little contentious," was "a good enough subject." Young L'Horme, "still a bit infantile," was "a good enough worker but somewhat quick-tempered."[84] Neither Jamet nor L'Horme, then, were barred from future stints at the mill. His father was trying to train choir boys, but rambunctious Jean-Pierre, the veteran of a tour de France, knew that boys would be boys. He was prepared to tolerate lively, skilled men with smart mouths, so long as they did not assault *him* with insults. A work force of Foucauldian docile bodies was surely beyond his ken.

Texier of Serrières had a part in a second incident, which began at the master's table. On October 3, 1787, the apprentice François Millot insulted some servants at a meal, knocked around some furnishings, and in the evening, beat a fellow worker. That night, Millot raised a racket in his bed. When one of the Montgolfiers arrived to impose order, Millot and Texier pursued him to the door, swearing and hooting all the way. Both men lost their jobs. There were limits to what Jean-Pierre tolerated, even in Vidalon's unsettled shops and household. Yet, Millot's certificat de congé, which recounted his mutiny, also noted that he could perform all the tasks around the vat and had "served loyally." After he gained some experience and grew up, Jean-Pierre concluded, the skilled Millot should be rehired.[85] Small wonder that Ecrevisse had journeyed to Lille in early 1781 in search of proper molds and docile workers.[86]

In April 1781, Vidalon's *patrons* had informed the absent Ecrevisse that "our [Hollander] beaters continue to roll along, we continue to be content with them."[87] Were they as satisfied with the ouvriers-élèves who accompanied the devices? Eager to lure Ecrevisse back to the mill, the Montgolfiers offered soothing words: "You will find, [we] hope, far fewer abuses [by the workers]

to reform than when you left." The Montgolfiers were struggling to install their new regime, and the conflict had cost them "several workers." "But I prefer," said one of the *patrons*, "to work only five vats at the moment and have only *bons sujets*." Other good hands, he predicted, would turn up.[88]

Four years later, in the midst of his pursuit of the runaway Brialon, one of the Montgolfiers maintained that the journeymen he had replaced in 1781 were "accustomed to debauchery and impunity."[89] Were François Millot, Texier of Annonay, and François Charbellet so different? In chapters 13 and 14 I provide quantitative measures of the attitudes and stability of the new men. Not long ago, these numbers would have been taken as yardsticks of labor commitment, a concept flavored with Cold War demarcations and side-taking. "Commitment," the sociologist W. E. Moore wrote, "involves both the performance of appropriate actions and the acceptance of the normative system that provides the rules and rationale."[90] Acceptance mattered little to Michel Foucault. From his perspective, the Enlightenment as a discourse had empowered the élite, including manufacturers, to impose its will and vision on the lesser sorts. All those indifferent to the logic of this "disciplinization" were treated as deviant.[91] Whether the working poor considered themselves as such is, of course, another question.

Jean-Pierre Montgolfier was unsure of the make-up of many of Vidalon's young hands. He was too astute, too experienced to label them unalloyed resisters or adapters. He knew that skillful, steady men had refractory sides. Deschaux was a good worker but "a bit unruly"; Jean Faya followed orders but was an "arguer"; and Girard, an "adroit, educated lad," capable of amounting to something if he mended his ways, lost his job because he was "insolent and spoiled" and "lacked respect."[92] Even Michel Chantier, one of Jean-Pierre's favorite hands, joined quick-tempered L'Horme in a drinking spree at a nearby mill in 1785.[93]

Every system of labor discipline spawns its own forms of resistance. Surely, shortcuts, surreptitious defiance, and moments of open challenge persisted under Vidalon's new regime. Yet, there was more afoot after the walkout of 1781, neatly symbolized in Michel Chantier's trek to another mill to imbibe. A new man who endured the Montgolfiers long enough to collect a 300-livre apprenticeship bonus, was he nonetheless seeking the approbation of the men of the "modes" in a nearby mill? Had he cemented unexpected links by drinking with them? Was he commenting with his feet and thirst on Vidalon's nouvel ordre, and if so, what does this flight by a stable Montgolfier hand tell us about the "commitment" of the bons élèves? Steven Kaplan has rightly observed that "we know extraordinarily little about the artisanal reception

and application of 'scientific' principles and procedures" of management and production.[94]

Herbert Gutman's famous query—what did the workers do with what was done to them?—can help here.[95] Certainly, the Montgolfiers' post-lockout work force remained playful, even if letting off steam was no longer tied to the "modes." While in an hôpital in 1785, the Vidalon hand Jean-Pierre Frappa threw eggshells at the servants.[96] Drinking bouts continued, and in July 1790, two workers kept horsing around behind the beaters after Marianne Montgolfier cautioned them to cease.[97] Like their predecessors, Vidalon's new hands seemed to have located the weak points in their bosses' new order. Already condemned as a "bad worker," Gache was fired because he claimed more posts of paper than he had actually made.[98] In 1786, Jean-Pierre also dismissed Olagne fils and Allemand for false counts.[99] How many other frauds succeeded, perhaps because Pierre's dependence on his large family for supervision still left too many hands without proper oversight?

Finally, workers and their families turned Vidalon policies to their own account. Journeymen on the tramp understood that possession of a congé could secure a day or two of food and shelter at the mill, even if they had no interest in toiling there. Knowing what the bosses liked to hear and see doubtless enabled some hands to travel from one Montgolfier mill to another. Families determined to situate their offspring, perhaps even an incorrigible youth, recognized that the Montgolfiers trusted lineage and physiognomy. Experienced Vidalon hands pursued premiums by giving notice. Workers eager to decamp told the Montgolfiers that they were needed at home to settle family business. There was an element of resistance in such activities, but even more, they demonstrated that Vidalon's hands had mastered the new order sufficiently to take advantage of it.

The Montgolfiers had their Hollander beaters and rationalized mill codes. They had double planted their work force and harvested skilled veterans as well as newcomers. They had separated skill from its traditional culture and could note in February 1789 that the journeyman Pierre Thibert "was working secretly to restore *la mode* in the mill."[100] They had deregulated their corner of the craft and were free to transform tools and techniques as they saw fit. But Vidalon's *patrons* still had to contend with the humanity of their skilled employees, their deceptions and reluctance, but also their desire to pile up bonus posts of inferior quality and petition for good food in exchange for good work. Skill was the resource that allowed Vidalon's hands to take advantage of what was done to them, all the more because they possessed precisely those skills that the Montgolfiers wanted and taught. Such skills were no mere

artifice, pumping up the pretensions of men who sweated in similar ways with similar materials. Production networks and the division of labor had not set paperworkers adrift among a flotsam and jetsam of journeymen with shared, rudimentary skills.[101] A particular set of skills and distinctive know-how permitted men and women to transform rags into paper. There was nothing else like it, which is why Pierre Montgolfier envied the Dutch effort to replace men with machines.

PART FOUR

Measuring Change

CHAPTER TWELVE

Technological Transfer

Desmarest and the Montgolfiers envisioned paper and profits enhanced by Hollander beaters and flexible, assiduous hands. By reforming one mill, they would light the way for an entire craft. They shared the zeal for improvement and enrichment of Wedgwood and Watt, and closer to home, the Johannots. All of these pioneer industrialists also shared similar problems. When the tilter at the Calder Ironworks came down with a fever, the firm could not fill a noted filemaker's order. The whole of Scotland, it seemed, lacked a competent replacement for one ailing man.[1] Far from unique, this tale illustrates the fragility of technological advance and the bottlenecks shaped by skill in a host of early modern European industries, papermaking included. Vidalon's Hollander beaters evidently worked best while Ecrevisse was in town.[2] In 1783, not long after this expert moved on, the abbé Alexandre-Charles Montgolfier implored Etienne, then in Paris, to secure a Flemish or Dutch foreman. It was an "absolute necessity," Alexandre-Charles confided, to acquire a "capable man" to supervise the beaters.[3]

Once the right man was forged or found, he needed the right tools. Here, too, delay was the norm. Ecrevisse, for instance, tested three combinations of copper and tin for the Montgolfiers in the ateliers of an Annonéen caster. Dissatisfied with the results, he turned to a fourth alloy and the shops of a Lyonnais founder.[4] Of course, the Montgolfiers' burdens had been eased by state aid, as well as their own wealth, reputation, and far-flung markets. For the petty producers out of Desmarest's sight, who struggled with marginal capital reserves and brief papermaking seasons, a flood, a fire, or a "famine" of rags threatened bankruptcy. In 1776, the vast majority of the approximately nine hundred vats that furnished paper in France were worked by such vulnerable men.[5] For most of these manufacturers, Hollander beaters, state assistance, and the creation of white printing and writing papers constituted a distant dream, if they could imagine it at all. This is the qualitative yardstick against which the spread of the beaters must be measured. Accordingly, the thump of 111 of these machines in fourteen généralités in 1789 was no mean feat.[6] Furthermore, in 1788, an official gathering samples for the national paper

collection remarked that several were "infinitely superior" to those acquired in 1776.[7] Desmarest's campaign and, perhaps, the Montgolfiers' pilot mill had triumphed, although the inspector would have been troubled by the fidelity of many producers to rotted rags and the use of the Hollanders to finish work started by the stamping hammers.

As late as 1812, according to a state survey, only 21 percent of French papermakers employed Hollander beaters.[8] Certain manufacturers continued to resist the device, and in two of the craft's centers, the Auvergne and Normandy, mallets alone macerated the pulp.[9] The Revolution also impeded the adoption of beaters, as the number of active vats fell to 598 in the year II (1793–94). Five years later, the minister of the interior, Nicolas-Louis-François de Neufchâteau, claimed that 1,061 vats in 745 mills turned out paper.[10] Establishing beaters in 21 percent of these mills, most of which housed one vat and hence were especially susceptible to the dislocations of the Revolution, constituted more than modest success.

The Montgolfiers were convinced that the construction of beaters in their mill would finally permit them to oust Dutch pro patria from the markets. In a letter that echoed the confidence of many, one of Vidalon's *patrons* boasted: "I am building a new establishment of beaters and Dutch-style shops to position myself to replace Dutch paper."[11] With slightly less bravado, the Montgolfiers explained to a customer in Bordeaux that they were enlarging their ateliers "in order to equal the paper of Holland."[12] A sense of urgency accompanied this optimism. Vidalon's *patrons* informed one correspondent that they "wish[ed] to press the construction of our beaters" and "[were] not wasting a moment."[13] Nevertheless, there were problems. "The building of the beater[s] is still giving us much difficulty," they wrote three months later, "and takes all of our time."[14] And there were costs: "I have not yet been able to fill your order," one of the Montgolfiers admitted to a Parisian client in 1780. He was busy installing "Dutch machines and techniques in [his] *fabrique*." All this would take four months more, but then his expanded mill would be able to satisfy the Parisian's commissions "promptly."[15] In the future, the Montgolfiers promised a customer in Montpellier, they would meet orders "with more exactitude and a more handsome paper."[16]

Producing paper that rivaled pro patria proved more nettlesome than the Montgolfiers anticipated. In April 1781, Pierre Montgolfier reported that "we have not yet been able to avoid *boutons*," an all-too-evident flaw.[17] Two years later, Jacques Anisson-Duperron of the Imprimerie royale complained about both the packing and quality of a shipment of Vidalon paper.[18] In June 1785, it was the turn of Etienne's old friend, the wallpaper manufacturer Jean-Baptiste Réveillon. He had recently received sixty-seven reams of Montgolfier

paper, of which a dozen were passable and only four close to former standards.[19] In 1787, Emeric David, the printer from Aix-en-Provence, toured the paper mills of Languedoc and Dauphiné. Several customers, he noted, thought that Montgolfier paper had declined in quality. He concluded that the Johannots' superfine paper was considerably better than Vidalon's, and he offered a brutal indictment of Montgolfier wares: purchases from Vidalon-le-Haut should be restricted to the inferior grades.[20]

What to make of these claims is difficult, especially since the Montgolfiers tirelessly advertised reams that matched the Dutch and stressed that they had supplanted the mills of Holland as the supplier of blue paper to the Royal Sugar Refinery in Montpellier.[21] The Johannots may have been right, to an extent: fame and flight had distracted Etienne Montgolfier from papermaking. While he was in Paris in 1783 and 1784, unsuccessfully pursuing further premiums for the mill and profits from ballooning, Vidalon's sales slumped. Despite their mechanical inclination, Etienne's and his father's austere temperaments may have also hindered the tinkering, the fine adjustments, that made improvement effective and profitable. Whereas David found Jean Johannot unceremonious, inventive, possessed of the "active and speculative spirit of a merchant," he observed that Etienne "want[ed] to be a scientist and to take the academic route."[22]

In part, it was the Montgolfiers' tendency to treat papermaking as an applied science that saddled them with a cost of innovation evaded by the Johannots: the recruitment and training of a youthful, new model work force. In 1783, the abbé Montgolfier wrote to Etienne that "our new workers labor no worse than [our] former hands, maybe even better." As we shall see, Jean-Pierre's assessments of individual workers seem to confirm his brother's impression. "But," the abbé continued, Vidalon's fresh, young workers "have more need of supervision and they are less clean." This last "quality," Alexandre-Charles knew, was "still the heart of papermaking."[23] Vidalon's ouvriers-élèves, however, had yet to learn this simple truth, along with the commonsense know-how once transmitted "de mâle à mâle"—know-how the new men had to learn from rules and regulations.[24] Small wonder that David claimed that "at Johannot's I think one sees greater cleanliness and a calmer order."[25]

That David also considered the Johannots' machinery "of better quality" is somewhat mystifying, since Ecrevisse oversaw the initial construction of beaters for both families.[26] Perhaps he was dazzled by their settings. Whereas the Johannots' new fabrique at Faya was three years in the making, Vidalon's machines were housed in an appendage of their standing enterprise—an "old French mill to which they added beaters," said Mathieu Johannot contemp-

tuously.[27] The payoff came in the market. While the Johannots enjoyed relatively steady gains, the Montgolfiers' sales climbed from 1778 to 1782, plunged for two years, recovered to pre-1783 levels by 1788, fell momentarily, and took off dramatically from 1790 through 1792.[28] In 1783, Etienne wrote that the mill at Montargis, another large-scale producer for upscale consumers, was marketing "very good" reams at relatively cheap prices. As a result, he was having "much trouble" disposing of his wares.[29] Yet, the Johannots and Montgolfiers both priced their papers high, which suggests that this mechanism did not account for the divergences in their markets. Nor did the foibles or persistence of the bons élèves and veterans employed by the Montgolfiers during the decade, if Jean-Pierre is to be believed.

From the "Law Book" of the Crowley Ironworks to rulebooks issued by the railroads, entrepreneurs have produced detailed, comprehensive programs of labor discipline. The fit between these schemes and the workers' daily toil and shopfloor conduct was inevitably inexact. But inexact in precisely what ways? This problem is magnified by the nature of the evidence scattered in the early factory masters' ledgers: it is weighted toward trouble, particularly trouble that prompted fines and dismissal. Finally, manufacturers petitioned the state for aid in controlling ungovernable hands, yet another depiction of workers as unruly and unsteady. The result may be a distorted account of the workers' actual performance, one that meshes too neatly with the entrepreneurs' perennial portrayals of incorrigible hands. Indeed, it is worth asking if the manufacturers would report or even recognize steady conduct and regular output when they witnessed it.

The Montgolfier archives offer two measures of their workers' everyday activities under the new order: the stability of male paperworkers during the 1780s and the productivity of Vidalon's vat crews at the turn of the century. Such statistical yardsticks, however, disclose little about the quality of the relations between Vidalon's masters and men. Here the Montgolfier collection provides an uncommon bounty: about seventy brief sketches of men and women who labored in the mill from 1784 through 1789. These character studies illuminate the attitudes and behavior that the Montgolfiers encountered and cultivated. Taken together, they suggest that Vidalon's *patrons* found more to cheer than condemn, certainly more to applaud than their public comments generally indicated.

CHAPTER THIRTEEN

Persistence

François Jamet was one of the novices who took the place of Vidalon's veterans in 1781. He put in fifty-six months at the mill during the 1780s, a remarkable figure in the footloose world of skilled men during the twilight of the Old Regime.[1] Nevertheless, Jamet, too, wandered. He left Vidalon-le-Haut in November 1785 to make his tour de France, only to return a year later with a certificat de congé from La Saône.[2]

To counter the wanderlust that surfaced in even the most steadfast of men, Augustin Montgolfier advised his family to build their new work force with "sturdy" local men.[3] His counsel paralleled the tactics of Ecrevisse, who had created a company of workers for a mill near Lille out of *"élèves"* from "the place itself or [its] environs."[4] Vidalon's *patrons* did not have to search far afield for young men willing to enter their shops. Of the ninety aspirants whose names appeared on the "roster of the workers trained by *le sieur* Montgolfier since he built the Hollander beaters," seventeen came from Annonay and four from Davézieux. Six candidates walked from Boulieu to the mill, as did four from Combe, four from Villevocance, three from Satillieu, and two from Serrières, all hamlets within an afternoon's stroll to Vidalon-le-Haut.[5] Doubtless, Vidalon's *patrons* realized, as Augustin had, that it was easier to transform young men into sedentary paperworkers when they were "*gens du pays* [locals] who have never left."[6]

Ecrevisse had selected mature men of thirty for apprenticeship at Esquermes, but Vidalon's *patrons* were crafting a labor force for the duration.[7] They chose younger hands, although for the most part, these men were well past the age of confirmation, the conventional start for apprentices in the trade. Perhaps the Montgolfiers reasoned that very young tyros lacked the size and self-assurance to stand up to the journeymen's association. Whatever the case, the age distribution of Vidalon's new workers (Table C) bore little resemblance to the layered pattern found in most paper mills. For example, during the year II, Courby Joubert's mill in Thiers employed a 63-year-old hand, two 50-year-old workers, a 43-year-old coucher, four men in their thirties, a 28-year-old layman, and six teenagers. (Of course, none of these men was of

draft age.) The vatman Jean Coste, who was fifty, may have sweated beside the layman Jean Coste, who was fifteen. At a second fabrique in Thiers, the governor Binet, a veteran of fifty years in the craft, supervised the preparation of the pulp that composed the sheets couched by 35-year-old Antoine Binet. The Montgolfiers frequently spoke of their green hands as *"jeunes élèves"*: these youths would learn the skills and rewards of papermaking from Vidalon's *patrons*, not their fathers.[8]

Labor turnover takes shape at the intersection of trade traditions and incentives, a particular entrepreneur's needs and resources, and the workers' desires and receptivity to the boss's inducements. Vidalon's *patrons* busily sorted through the young men of their region in search of paperworkers and then attempted to stabilize those they had chosen. Not surprisingly, many half-trained aspirants quickly fell by the wayside. It is possible to identify forty-three paperworkers and governors at work in the Montgolfiers' principal mill in April 1784. (Etienne Montgolfier estimated that he employed a total of 150 people there in 1786.) Thirty-five of these skilled men took up the craft at Vidalon-le-Haut from 1780 through March 1784. Table D depicts their tenure in the mill: during the 1780s, nineteen of these men spent five or more years in its shops, although like Jamet, they also took their leaves. At the close of the decade, ten of these durable hands still labored there.[9] In light of the hazards of the craft, from rheumatism and pneumonia to losing one's "shake," such persistence was remarkable. Similar stability was not unknown in the shops of contemporary producers who combined substantial capital and skill. Of eighty male hands who toiled at Oberkampf's calico-printing works at Essonnes from 1760 to 1820, just under half lasted six years or more.[10] In England, Boulton and Watt also built up "a team of skilled mechanics." They accepted pauper apprentices, men in related skills, and like Vidalon's *patrons*, relied on long-term bindings.[11] More typical, of course, was the glazier Jacques-Louis Ménétra, who switched Parisian bosses six times in less than two years.[12] And the Montgolfiers' stable core of home-grown hands was uncommon in papermaking as well. About the journeymen of Thiers one subdélégué observed: "Since most of these workers are not domiciled and are not attached more to one *pays* than to another, they decamp on the first whim that comes to them."[13]

Did the threat of the men of the "modes" shackle the Montgolfiers' élèves, however well trained, to Vidalon-le-Haut? Valençon, it will be recalled, landed work in a nearby mill, and Jamet made his tour de France. Jean Charbellet of Davézieux, who caught on with the Montgolfiers in 1782, left a year later. Jean-Pierre readily rehired him in 1786 when he returned with a written discharge from a fabricant in Uzès. Seven months later, he was gone again.[14] Such itin-

eraries suggest that menacing journeymen did not keep the Montgolfiers' new hands at Vidalon-le-Haut. They also indicate that the Montgolfiers were prepared to tolerate a certain flightiness, even in good hands like Jamet.

While continuing to train apprentices throughout the 1780s, the Montgolfiers also replenished Vidalon's work force with journeymen paperworkers and governors, the vast majority of whom earned their spots with certificats de congé. These men followed several paths to Annonay. Their treks offer the tracings of a regional labor market, with little representation from the large paper mills of the Parisian basin. One route ran parallel to the Rhône, linking Vidalon-le-Haut to the fabriques of Vaucluse and the Comtat, while a second trail led through Dauphiné to Savoy. A much-traveled passage brought workers to the Montgolfiers from the great papermaking centers in and around Ambert and Thiers of the Auvergne, as well as the small mills along the way.[15]

Although Vidalon's *patrons* offered regular employment to these old hands, the hiring of many during seasonal surges of work or in the aftermath of dismissals colored their arrivals with a temporary cast. Judging by his congé, sickly Jean-Joseph Micolon endured a 124-day odyssey before he landed a job at the mill.[16] As Table E demonstrates, few of the journeymen's treks lasted remotely as long. Of the thirty-three trips that can be measured (on the basis of the dates inscribed on the itinerants' discharges), fourteen took nine days or less, and only six took a month or more. Mercier covered the 35 kilometers from Tence to Annonay in two days; Joubert spent six.[17] With his wife and three children in tow, Louis Filhat managed the 66 kilometers from Rives to Annonay in two days.[18] Yet, fifteen of thirty strolling men completed only 1–7 kilometers per day on their way to Vidalon-le-Haut.[19] Evidently, they had sweated in the shops of fabricants who failed to mark their congés, or discovered other means of survival. In any event, these workers exhibited no great hurry to find work with the Montgolfiers.

Perhaps these men took their time because the road, as dangerous and cold as it often was, also offered the prospect of encountering familiar faces. Consider the connections among the paperworkers of the district of Thermopylae (formerly Saint-Marcellin) during the year II. The region housed three paper mills: Bardon's fabrique in Izeron, Blanchet's in Rives, and Jubié and Cie in La Saône. Each of these mills had lost one or more hands to the Montgolfiers during the 1780s. Of the twelve men and women laboring at La Saône during the height of the Terror, four had once toiled at Vidalon-le-Haut. Two of the ten workers at Bardon's mill had put in time with the Montgolfiers. Among the eleven male workers at Rives, three (and possibly a fourth) had sweated in Vidalon's shops. Altogether, 27 percent of the district's paper-

workers had spent part of their lives laboring for the Montgolfiers. Moreover, Chanteloube, Charbellet, and Duranton, established names in papermaking along the Deûme, also turned up in the mills of Saint-Marcellin. Such links naturally made tramping less forbidding.[20]

Blaise Duranton took to the road with a good chance of meeting kin wherever he ventured. At the close of the Old Regime, Durantons could be found across the papermaking landscape of southeastern France. Apparently, the family's roots were in Marsac, near Ambert, in the Auvergne. By 1784, Blaise had made his way to the family-supervised mill in Tence, a village 35 kilometers southwest of Annonay. This humble enterprise was part of the Montgolfier matrix and, like their other small mills, seems to have served as a feeder for the big mill. Blaise soon gravitated there, joining his wife and Antoine Duranton, the former soldier who also carried a certificat de congé from Tence.[21] In January 1785, Claude Duranton, who was either Antoine's brother or nephew, left Tence and sought work at Vidalon-le-Haut.[22] Four months later, Jean-Pierre fired Blaise.[23] His trail is again visible in 1793, when he surfaced at Blanchet's mill in Rives.[24] Durantons from Marsac made paper in the Johannots' shops, at several mills in Thiers, and at Vidalon-le-Haut during the Terror.[25] Small wonder that Vidalon's *patrons* selected their new men largely from families innocent of the craft.

Table F depicts the persistence of the skilled hands hired on the strength of their congés by the Montgolfiers from April 1784 through December 1789. For the most part, their short stints remind us why officials and producers alike labeled them "gadabouts" (*coureurs*): about half of these men lasted at the mill for three months or less, and only a fifth remained there for more than one year.[26] Still, once a man got a firm start in the trade, he tended to stay with it: during the year II, for instance, the fifty veterans and apprentices who toiled at the large Johannot mill had been making paper for an average of twenty and a half years.[27] Apparently, this commitment was accomplished without much loyalty to any particular mill. How had this paradox of stability in the craft and ceaseless roaming come about?

Although manufacturers and the state centered on "feckless" journeymen, much of the answer lies in the mills the workers abandoned. When Jean-Baptiste Pons arrived at Vidalon-le-Haut in 1785, he showed Jean-Pierre a certificat de congé endorsed by one Bernard, the master of a marginal mill.[28] Thirteen years earlier, Louis Gentil had held the lease at this fabrique, whose sole vat, often idled by scarce water, had papermaking seasons of only four months.[29] A mill in Uzès was Odoin's last stop before he put in two and a half months at Vidalon-le-Haut.[30] In 1772, an enquête doubted whether this one-vat concern would ever produce much paper.[31] Winter cold "frequently" dis-

rupted work at the two-vat enterprise in Tence, which provided Vidalon-le-Haut with four hands whose stints lasted one day, three months, seven months, and eight months.[32] And some Dauphinois producers, according to their anonymous chronicler, had to forgo journeymen altogether. "Dying of hunger in their shops," they performed all the tasks themselves, turning out a few reams of "common paper" to cover their needs and the purchase of some cast-off linen.[33] While the lure of the road seduced some paperworkers, the irregular rhythms of production at many mills gave others no choice. Such men could have little "expectation of security in employment" and little trust in those manufacturers who promised it.[34] Workers like Odoin and Blaise Duranton filled the periphery of Vidalon's labor force, plugging holes opened by illness and indiscipline among the jeunes élèves, taking jobs created by flows in the millraces, and then moving on.

The Montgolfiers replaced a labor market dominated by journeymen and their culture with an internal labor market that provided a stable core of employees. Orbiting around these regular men was a shifting constellation of changeable newcomers, short-time itinerants, and a handful of hardy journeymen, some of whom evidently picked up the Montgolfiers' ways in their lesser mills. There was reason for Vidalon's *patrons* to cling to their new order through the ebbs and flows of profits. The journeyman who lasted a year at the Société typographique de Neuchâtel, a printing shop just over the French border, was known as an ancien.[35] A considerable number of Vidalon's new men merited similar designation, thereby easing the Montgolfiers' task of coordinating the work of skilled men.

CHAPTER FOURTEEN

Attitudes

Vidalon's *patrons* wanted men who would play their roles as precisely as they were drawn, without comment or complaint. Like Robert Darnton's Swiss printers, they sought diligent, sober hands.[1] Like Michael Sonenscher's Parisian master craftsmen, they prized "adaptability." Sonenscher's masters and manufacturers orchestrated out-of-doors fabriques, alleys of enterprise where skills and raw materials had diffused through a wide assortment of trades. In this setting, the traditional craft worker, tethered to a particular, quality product, was less valuable than the flexible quick study, capable of applying his know-how to a broad range of semifinished goods.[2] For the Montgolfiers, however, the adaptable man was the son of a vigneron or day worker willing to apply himself to the mastery of their art—the fabrication of quality reams for discriminating customers.

In their correspondence with officials and among themselves, the Montgolfiers echoed their era's received wisdom about paperworkers in particular and journeymen in general. They were feckless, footloose, turbulent, and unreliable. They were thinly skilled, and drink spoiled even the best worker and his work. "They work badly and cheat the public," said the Dauphinois reporter.[3] Augustin Montgolfier explained in 1781 that journeymen paperworkers constituted a "Race" composed largely "of true hacks worth no more than apprentices of six months."[4] Years earlier, however, when his own workers were seized by the "frenzy" of débauche, pronounced fines against him, and banned his mill, he bought them off.[5] Evidently, he wished to lose neither production nor his skilled hands.

Jean-Pierre's appraisals of the men who toiled at Vidalon-le-Haut from April 1784 through December 1789 certainly deviated from his family's usual public and private claims. He was evaluating for rehire workers who had just quit or been sacked. Thus his impressions were candid and immediate, and all the more surprising in light of the sullen and stormy ends of many stints. He would eagerly take back two men; five more were "good to rehire"; nine more should be signed on should the opportunity arise; one could be rehired in the absence of better; three should only be brought on in pressing cir-

cumstances; and only two should never again sweat in the mill.[6] As Table G demonstrates, Jean-Pierre found two excellent or good workers, or men who performed their duties well enough, for every mediocre or poor paper-worker. Assiduousness mattered: Brunel was "accustomed to doing nothing" and "something of a goldbrick."[7] So did skill: Roderic was a "good enough coucher and layman"; Chaumier was a "passable vatman" and "not [a] bad coucher"; Barré was a "good enough layman"; and Odoin was a reasonably good vatman.[8] Jean Charbellet was "adroit," but Grenier was "a very mediocre worker."[9] Faye cut the pulp well and, better yet, "[did] what one told him."[10] Disposition always caught Jean-Pierre's eye: Frappa took "enough care with his work," but Thibert was no more than an acceptable worker when watched and something less when Jean-Pierre failed to look over his shoulder.[11]

Ever mindful of the "modes," the Montgolfiers feared the stiff-necked, contentious man most. Neither loyalty nor obedience could be expected of him; he was incapable of adjusting to a carefully designed factory regime. Consequently, Jean-Pierre worried that the "argumentative" Nicolas Gonein would never be a "good subject."[12] But the "rather mild" Renard was worthy of a second chance if he returned. Peschi, a gentle, quiet man "full of religion," had no flaws other than his advanced years. Doire and Jelbi, both of whom were tranquil, received grand praise: they were *"sage,"* that is, prudent.[13] In sum, Jean-Pierre characterized twenty male members of Vidalon's post-walkout work force as either calm, docile, mild, polite, tranquil, or "never insolent." He encountered only nine men who, when free of the influence of wine, were contentious, disrespectful, headstrong, insolent, or mutinous. Four others turned disputatious after they had drained a jug or two.[14] These figures suggest something about the Montgolfiers' aggressive winnowing. They also imply that Jean-Pierre turned the other cheek repeatedly before branding a man insubordinate. After all, one of Vidalon's *patrons* still found it necessary to demand that "the workers will behave with the decency and the respect that they owe to their master without swearing, insulting, or disputing in the shops or at the table."[15]

Ironically, Jean-Pierre did not find tight-lipped men reassuring. He fretted that a follower like Roux might become attached to "sly" Jean Charbellet, "suspicious" Le Bon, or the conspirator Thibert.[16] Already concerned about the "suspicious" families of Frappa and his wife, he was troubled that this long-term Vidalon hand had a way of putting his comrades under his thumb.[17] Jean-Pierre dreaded such cunning, yet valued intelligence. He poked fun at Michel Duranton, a "good worker" who was "stupid" and "a little behind the times."[18] Whereas one hand had "enough education," a second was an Auvergnat "of uncommon stupidity."[19] Marie-Anne Chatron dit Riban was quick

and could read, write, and figure. Here was a bright woman, capable of taking advantage of Vidalon's new order. Jean-Pierre recommended her for rehire but portrayed her as a sharp minx who could do much harm should she turn out badly.[20]

Even the Montgolfiers' comparatively restrained new hands, as we have seen, found occasion to bicker and square off, and had to be urged "to live tranquilly and without dispute."[21] Sobriety, Vidalon's *patrons* believed, was a cardinal virtue. Jean-Pierre clucked disapprovingly that Perrier was "playful when he [went] to town."[22] Waggish women particularly aroused Jean-Pierre: Claire Maret was "a bit playful, provoking the workers," and Claire Beraud was a girl "without morals, immodest, and capable of corrupting others."[23] Such women might hide the vatman's wine or join the men in one of Annonay's taverns.

Wine, Pierre Montgolfier scoffed, "embolden[ed] [*monte l'imagination*]" the workers. Worse yet, the "greater the number [of workers], the more they [were] violent in their deliberations."[24] Alcohol turned valuable workers into insubordinate hands: Englement, a "good worker," became a "quibbler" when he imbibed, and Julien Bourgade, who generally behaved and produced well enough, was "a little insolent when he drank."[25] Drink encouraged craftiness: Damien Le Bon always found "excuses to drink."[26] How could a journeyman perform the delicate art of papermaking if, like Pracrau, he was a "determined drunkard, stupefying himself with wine"?[27] Insobriety also undermined family harmony and discipline. Chatron dit Riban was an "incorrigible profligate, besotted by wine," who battled with his wife and mistreated his children. His wife, who drank heavily herself, was a "subject even worse than [her] husband."[28] Above all, to Augustin Montgolfier's horror, drink encouraged solidarity among the workers: "never suffer any pump stand [*buvette*] around the vats," he warned Vidalon's *patrons*, "not even the *pot de grand;* you know how wine brings together the men whom you have an interest to separate and to hold apart as much as you can."[29] Accordingly, the man caught drinking in the mill faced a stiffer fine than his sodden sister or wife.[30]

Even capable men, the Montgolfiers recognized, drank immoderately. Jean-Pierre described six men who left the mill from April 1784 through 1789 as drunkards. One of the six was also dazed by wine, as was a seventh man. An eighth hand was a little too fond of his wine and five more were "debauched," unruly and prone to fits of drunken idleness. Perhaps the four men possessed by pot-valor worried Jean-Pierre most. So he gratefully acknowledged that four of Vidalon's workers were not "debauched" and tolerated men like Grangeon, a moderately quiet, skillful sot.[31] A tranquil drunk did not cause many ripples in the shops and residences.

"Sans raison" when he drank, the vatman Joseph Etienne lost his job at Vidalon-le-Haut on March 1, 1785.[32] Etienne began his bumpy career with the Montgolfiers in January 1784.[33] He was fired and rehired nine and a half months later. By February 1785, things appeared to be going well between Etienne and the Montgolfiers; he had even stepped up to a vatman's station. But he evidently went on too many binges—usually twice a month—and forfeited his job on March 1.[34] Jean-Pierre took Etienne back a day later, with the promise that he would discharge the vatman the next time he caught him drinking.[35] Three days later, an exasperated Jean-Pierre wrote, "Etienne, who has not ceased to besot himself since the second of the month, has been booted out." Out of "consideration for [Etienne's] wife," Jean-Pierre relented once again and rehired the skilled man on April 11. He also informed Etienne once again that should he "return to drunkenness," he would surely be shown the door.[36]

On August 21, 1785, Etienne reappeared in Jean-Pierre's accounts. He left that day with a certificat de congé that depicted him as an "incorrigible drunkard." In a note to himself, Jean-Pierre observed that Etienne was insolent when drunk and completing his self-destruction with staggering quantities of tobacco.[37] On November 27, Etienne was back at Vidalon-le-Haut. Eleven months later, he abandoned the Montgolfiers because Jean-Pierre no longer permitted him to work the molds. Etienne was no longer sure of his weights, Jean-Pierre reported, but had been fairly well behaved during his last stint.[38] Perhaps Etienne was accurate only when the monotony of promenading fresh sheets was broken by regular drinking bouts. He was willing to drink himself out of a job but not to step down to the coucher's felts; Etienne never returned to Vidalon-le-Haut. As far as Jean-Pierre was concerned, his skill was gone, and so was he.

His replacements, the Montgolfiers believed, would exceed his skill and forswear the vices that had eroded it. Theirs was a teaching regime, geared to transfer their version of refined skills and useful habits. They would ensure that young Marie and Claudine Chatron dit Riban, "having the necessary dispositions," became good workers.[39] They would winnow Vidalon's ranks so that Odoin, who would be a "well-behaved lad" among "prudent" companions, had the chance.[40] They would instruct "loyal" Pourra, who was "not lacking in intelligence," in the virtues of self-interest.[41] And they would shield men like Jeannot Pigeon, who had good instincts but a "bad education," from Joseph Etienne's destiny, that is, if they did not have to fire him too.[42] Their management and their rule had merged, with no room for the journeymen's custom.

CHAPTER FIFTEEN

Productivity

The drinking sprees, cockfights, and charivaris that disrupted the workday in early industrial shops are better known than the workday itself. In public and private, pioneer manufacturers in England and France sang in one key: their workers toiled irregularly, in large measure because they brought habits learned in the field and cottage into their mills.[1] Even where machines governed work, entrepreneurs had to struggle mightily to separate labor and leisure. Threadbare incentives did not help. It took the steady pulse of the steam engine, plus bells and whistles, fines and schoolings, and perhaps a widened array of goods within the workers' reach to overcome the legacies of croft and craft.[2] Meanwhile, customs such as Saint (Blue) Monday mocked the steady intensity that even the most notable producers could barely conceive.[3]

Yet, the occasional flare-ups recorded in an entrepreneur's daybook, his disgust with workers drawn into the street by a mime or a hanging, and his open disparagement of feckless hands reveal little about routine activity in his shops. Above all, the particulars are necessary: the scale of the enterprise, the resources of the manufacturer, the extent of his market, the virtues (and drawbacks) of his location, the availability of raw materials, the division (and synchronization) of labor, the level of skill, and the craft heritage of productivity. Like most Old Regime industries, papermaking was punctuated by spontaneous feastings and seasonal shutdowns, not to mention its particular susceptibility to the smuggling of old linen. Still, its reliance on rotting rags and connected tasks meant that both masters and men knew that time was money. Equally, the daily quota of twenty posts of paper specified by Desmarest in 1788 was precisely the figure set by the English anti-combination legislation of 1796.[4] Was this standard honored mainly in the breach? Or was papermaking's heritage a subtle blend of long nights of labor interrupted by respites and walkouts, all of which was deemed mercurial by hostile bosses and uninitiated observers?

At the Lana mill in Docelles in 1827, when the modern papermaking machine had not yet become a visible challenge to the journeymen, all five vats turned out paper on nearly three hundred days.[5] Forty years earlier, Vidalon's

patrons calculated the value of their workers' wages and food on the basis of an identical schedule. Of course, these examples are not intended to suggest that coopers, carpenters, fenmen, and fishermen endured uniform workdays throughout the year. Instead, as David Landes once observed, had domestic production been capital-intensive, as hand papermaking was, the merchants would have hired someone to go round early in the morning to awaken the spinners and weavers.

Just as the Montgolfiers secured a core of persistent, somewhat subdued hands in a tumultuous trade, they obtained surprisingly constant output. Perhaps, however, the surprise is mostly ours. We have grown accustomed to associating consistent production with the sputter and belch of machines—a wonderful irony because many of these early devices broke easily and often. A rather simple instrument, Hollander beaters were less likely to seize and shudder, and thus they fit neatly with Vidalon's long papermaking season, fixed workloads, and set workday.

The Montgolfiers' tireless experiments with technology and discipline extended to the keeping of accounts. In December 1798, they introduced a new output register (*livre de fabrication*). Each vertical column depicted two weeks' work at one of the six vats. (The dislocations of the Revolution had reduced the number of active production units at the mill to five, and occasionally six.) The Montgolfiers placed the names of the vat crew, plus an apprentice, at the head of each column. Each horizontal line represented a day. In the boxes blocked out by this design, Vidalon's *patrons* entered the trade name, weight, and color of the paper worked by the team, its output, and the identities and stations of any substitutes.[6]

This seemingly straightforward system, a rational complement to a rationalized regime, was complicated by remuneration for breach days. Common in both the French and English trade, this customary payment evidently stemmed from the papermaker's responsibility to provide work for his charges. The producers of Kent, one of the centers of the English craft, "agreed to find work for their men for six days per week and 'when short of water to find them other employment equivalent thereto.'"[7] In the Vosges, paperworkers at one mill received a half day's wage when a drought or a freeze idled them. But when the problem was scarce rags, a man-made deficiency, the journeymen were entitled to a full day's pay.[8] Such compensation, even makework, was unusual in an era when shutdowns generally drove both the skilled and unskilled to the road. But rags rotting in the pourrissoir, the promise of seasonal rains, and the journeymen's effort to keep their ranks thin led papermakers to cling to their skilled hands. As one inspector of manufactures explained, "Want of a single [paperworker] halts the work of three."[9] So manufacturers coughed up

a few sous and hoped for the speedy end of a drought or the quick repair of a cracked vat—and, as soon as a campagne was filled or a papermaking season reached its definite conclusion, sent the journeymen packing.

As already mentioned, Vidalon's paperworkers earned a fixed, reduced reward for the odd jobs they performed on breach days. When the vatman ran out of pulp in the midst of production, his team was credited with a partial breach day. Thus a half breach day plus the fabrication of a dozen posts of *bâtard fin* occasionally added up to more than a full day, reminding us of the always difficult marriage of piece rates and time rates. Factor in the premiums for meeting and exceeding quotas, and the Montgolfiers faced a bookkeeping nightmare of both their own and the craft's making.

"Redundancy or Deficiency of Water, want of Materials, Intervention of Holydays, and other Contingencies" resulted in "the many Interruptions" that afflicted England's papermakers.[10] Accordingly, Lalande's claim of three hundred days of work per year was certainly inflated for many mills, but not by all that much at Vidalon-le-Haut.[11] As Table H illustrates, in 1799, 1800, 1801, 1804, and 1805, Vidalon's five vat crews regularly spent about 250 days per year making paper as well as another 35 days or so refurbishing tools and canals. Even in the difficult years of 1802 and 1803, the vat crews turned out paper on approximately 228 and 210 days, respectively. Taken together, these figures reflect the market and technical circumstances of the Montgolfiers' nouvel ordre and help explain their desire to train bons élèves for the years.

Throughout the eighteenth century, masters and manufacturers in many trades, and papermakers in particular, lamented the large number of feast days enjoyed by their workers.[12] George Rudé maintained that Old Regime workmen, from the laborer in Réveillon's mill to journeymen locksmiths and masons, had 111 unpaid holidays per year.[13] Contested claims, chronic underemployment, and staggering hours in summer sunlight render Rudé's reckoning little more than an assumption. For their part, the journeymen paperworkers of Protestant England had twenty-four and a half holidays during the 1790s.[14]

At Vidalon-le-Haut, the Shrovetide Days, which had figured prominently in the Montgolfiers' accounts during the 1780s, were regular workdays by the turn of the century. As Table I reveals, their hands enjoyed an average of fifteen full and three half-holidays from 1799 through 1802, and about half that number beginning in 1803. These spare holiday calendars were remarkable but not unique; in fact, they were close matches for the feast-day schedules in the paper mills of the Paris basin and Angoulême, the latter in 1827.[15] Eager to protect the festive life sanctioned by their "modes," perhaps some journeymen paperworkers gave ground more easily on the conventional liturgical

cycle. Whatever the case, lean feast-day schedules certainly enhanced the lengthy papermaking seasons at the Montgolfiers' principal mill.

Although Lalande declared that "paper-making [was] carried on throughout the year," French production generally slowed as spring torrents turned into summer trickles.[16] Table J depicts the monthly rhythm of work at Vidalon-le-Haut. In effect, the Montgolfiers organized production in two papermaking seasons. One stretched from mid-October through May, from revitalizing fall rains through cool winters and April downpours, which once again quickened the millwheel. During the seven-month span from November through May, the five principal vat crews averaged as many as 24.1 days of papermaking in March and no less than February's 21.5. Remarkably, the teams also averaged less than one breach day both in April and in May.

In June, the average number of papermaking days slipped to 18.5. Annonay's hot, dry summers resulted in a low of 14.1 papermaking days in September. Autumn showers reanimated the millwheel, as the jump in October workdays demonstrated. Heavy rains and muddied millraces, however, could also disrupt fall production; in September 1801, for instance, an inundation compelled the Montgolfiers to halt the pulping. Still, early autumn and summer output distinguished good years from bad. In robust years, such as 1799, 1801, and 1804, production was largely undisturbed in three of the four months from July through October. In thin years, such as 1802–3, it was limited in each of these months. Even so, during the seven years under study, September, the leanest month, averaged only eight fewer papermaking days than most of the flush months—a far cry from the three months of every year that Crozier of Vals and Filliat of Antraigues shut down.[17]

Although the Montgolfiers kept their books by the Revolutionary calendar, nothing suggests that they tried to convert the work week from six days to the *décade* (ten-day week). Evidently, Vidalon's *patrons* felt no pressure to reduce the number of Blue Mondays; in fact, Saint Monday was just another workday at their mill (see graph in the appendix). Output peaked on Wednesday in 1799 and 1801, and on Tuesday and Saturday in 1804 and 1805. In 1802–3, production reached its peak on Saturday alone. Saturday's yield exceeded that of Monday annually, although never by more than 9 percent. The frequent assignment of new formats on Monday morning accounted for some of this pattern, as did pressure by the *patrons* to use up the last of the perishing pulp before Sunday.

In 1803, a meager year, and in 1804, a fat year, Vidalon's vat crews met their quotas on 54.3 and 53.7 percent, respectively, of all the full papermaking days they worked (see Table K). When they failed, it was most often by one post of paper, that is, forty-five minutes of work. Generally, these shortfalls occurred

on Monday, when the teams changed and cleaned the molds. They surpassed expectations on just under 4 percent of their complete workdays in both years, usually to consume the last of the pulp before moving on to another sort of paper. Evidently, Vidalon's regular workdays coaxed consistent output rather than extraordinary achievement from the men, just as the Montgolfiers intended. As Desmarest had observed, enterprising papermakers asked no more from their workers than to complete the *tâche ordinaire*. Anything more courted disaster, especially in the Montgolfiers' tony markets.

In 1761, the vatman Vissier had a good year: Pierre Montgolfier credited him with 274 workdays and a share in 5,766 posts of paper. Exceptional weather loomed large in Vissier's success: he worked twenty-three days in August and another twenty-two in September.[18] His output was in the same range as that of the vatmen in the hearty year of 1801, when five crews (and, for a month and a half, a sixth) fashioned 28,795 posts. (It should be noted, however, that Vissier concentrated on fine, lightweight papers. In 1801, the vatmen labored on a considerable amount of heavier writing and printing paper, at quotas of two to four posts less per day than Vissier's lightweight wares.) Still, there is no escaping the conclusion that the pace of work at fin-de-siècle Vidalon-le-Haut did not differ dramatically from that of 1761. Nor is this startling. After all, Hollander beaters or not, the basic division of labor in the craft remained largely unchanged; the beaters themselves, associated with quality reams and deeper molds, regularized and prolonged the vat crew's labor rather than hastening it. If anything, Dutch masters reputedly worked their hands more slowly than their French counterparts. The Montgolfiers wanted more product, but even more, they longed for reams that rivaled pro patria.

If brief papermaking seasons and the "modes" did not signal a craft heritage of steady intensity, there were legacies of a different sort. The master's table carved workdays into units of time and output. Breach day payments and makeshift work reinforced the contours of conventional workdays, as did widely diffused quotas. Everywhere, the vat crew was expected to gather at the press, a custom Vidalon's *patrons* transformed into a lesson: fines paid by hands who failed to pull the bar filled the pockets of those who did.[19] Even overtime hours schooled workers in the borders and rewards of an ordinary workday.[20] Already in seventeenth-century Italy, "the average daily production of one vat did not exceed a maximum of 4,500 sheets," a measure quite similar to that of early nineteenth-century Vidalon-le-Haut.[21] That ceiling was about to give way, but not through the efforts of effective managers, who dared not sacrifice quality to challenge it. It took the papermaking machine, patented in 1799, to set a new pace.

CHAPTER SIXTEEN

The Hierarchy of Vats

The stick certainly outweighed the carrot in early systems of industrial labor discipline. In mills of every sort, fines, corporal punishment, dismissal, and the threat of discharge took precedence over premiums and promotion.[1] One result is that historians know more about the design and purpose of bounties and advancement than about their efficacy. A precise account of the workers' response to each inducement offered by Vidalon's *patrons* is also beyond reach. It is possible, however, to chart the transit of individual hands up and down the ladder of production units and to consider, as the Montgolfiers themselves did, the flow of work and assignment as a sturdy incentive.

Just before the lockout, Augustin Montgolfier advised his family to create a hierarchy of vats in their principal mill. Under his plan, every worker would rise or fall in accord with his work and conduct, and every dipping vat would be associated with a particular grade of paper—*surfin* (superfine), *fin* (fine), *moyen* (middling), *bulle* (blistered), or *trasse* (bottom). Augustin claimed that Vidalon's *patrons* would obtain more than improved reams: his system would foster a healthy jealousy among the men sweating at neighboring vats and might even yield an informer.[2] His counsel came after the fact. In July 1777, Vidalon's *patrons* paid 20 livres to the team *"à la cuve du superfin."*[3] Of course, the rest of the Montgolfiers did not have to share Augustin's analysis to entrust the most valuable paper to the most skillful hands. Nevertheless, the family preserved this visible measure of their esteem and the vat crews that developed within it.

Twenty years before the walkout, Pierre Montgolfier generally assigned fine paper to the vatman Vissier, and bulle and moyen to Jarsaillon.[4] In 1799, Pierre's heirs supervised the labor of five teams, the last of which shifted between the fifth and sixth vats. As Table L illustrates, Vidalon's *patrons* entrusted the bulk of their heavyweight, superfine paper, which paid the most, to the hands of the first vat. The crews at the second, third, and fourth vat also fashioned superfine reams, but most often they weighed and paid less. On December 30, 1798, for instance, the first team began work on *raisin surfin* at forty pounds per ream, an exceptionally heavy paper. The third unit

was already busy with *écu surfin* at fourteen pounds, and the fourth with *coquille surfine* at sixteen.[5] On the basis of the Montgolfiers' *tarif* (output and reward schedule) of 1789, the vatman making *raisin* would have earned 5 sous more for a complete workday than either of his fellows—a considerable incentive.[6]

A great tide of relatively lightweight fine and superfine paper washed away most distinctions among the middle vats. This dilution did not extend to the fifth unit, however, which produced all of Vidalon's trasse, almost all of its bulle, and much of its moyen in 1799. Ironically, the Montgolfiers' focus on fine wares left little room for a neat hierarchy of vats, except at the edges.

When the Montgolfiers assembled an effective vat crew, they labored to keep it together.[7] They dispatched whole teams to functional vats when their regular equipment failed and worked them "day and night."[8] In July 1802, the fourth and fifth crews made paper at the sixth vat in overlapping shifts. With the Hollander beaters still down in September, Vidalon's *patrons* diverted the first team to the sixth dipping vat, where it labored alone. When making paper proved impossible, the Montgolfiers declared a breach day and sent entire units to muck the race or refurbish tools. Rotating a cohesive team, whether to a working vat or makeshift work, maintained its identity, and probably kept skilled hands from the road.

The careers of the vat crews mirrored their standing in the hierarchy. From January 1799 through November 1805, only six men occupied the skilled stations at the *première cuve*. Périgord made the paper, Châtagnier couched it, and Zacharie Filhiat prepared it for the press for at least thirty-three consecutive months. After a short stint by a second vatman, Barthélemy Bourgade took over the molds in January 1802. The crew then remained intact, except for two weeks, through November 1805. Clearly, the Montgolfiers selected stable, skilled men for the first vat: Périgord, Châtagnier, and Filhiat had all started at the mill before the Revolution. Yet, Bourgade had not served around a Montgolfier vat during the three years before he obtained his elevated station. Skill mattered.

At the other end of the spectrum, the Montgolfiers rummaged through the ever-changing fifth vat crew for the spare parts necessary elsewhere. Charnier, for example, enjoyed some time at the third vat in 1804 before returning to his familiar station on the bottom rung. A month-long assignment as the coucher at the fifth provided the apprentice Gardon with his first toehold on the ladder of skilled posts. And the coucher Châtagnier tried his hand as a vatman at the fifth when water problems idled the première cuve. The fifth vat hosted forty-three men in regular, skilled positions from January 1799 through November 1805, the greatest turnover of all. Still, as a storehouse of replace-

ments, a testing and training site for the untried, and a harbor for prized men, the last vat enabled the Montgolfiers to maintain their production hierarchy.

As Vidalon's *patrons* balanced the flow of work and assignment, they also determined the trajectories of individual hands. As we have seen, seniority alone did not propel a man up the ladder of rewards. Jamet and Frappa sweated for years with Vidalon's felts without ascending to the molds. But Artaud spent thirty days as a coucher, put in a short time as the vatman of the fifth team, and quickly succeeded to the same role at the fourth. Small wonder that Vidalon's codes emphasized the Montgolfiers' latitude in assigning tasks and stations, and banned interference by workers—the envious as well as the fraternal—in the affairs of their fellows.

Vidalon's fin-de-siècle records offer nothing to match Jean-Pierre's earlier characterologies. Without such appraisals of skill and comportment, one can only speculate about the mix of ambition and steadiness that seemingly characterized François Bourgade. A coucher at the sixth vat in 1800, he moved up to the same post at the third a year later, and finally secured the felts at the second in 1802, an assignment he still held in November 1805. For his part, Barbot had a one-month taste of life as a coucher in December 1802; otherwise, he lingered as a layman, almost always at the fourth vat, from September 1800 through November 1805. Was he frustrated, indifferent, incapable of mastering the coucher's and vatman's more lucrative skills? However inviting their incentives were, the Montgolfiers were still constrained by the human vagaries of their skilled employees.

Jean-Pierre often noted that a young man had been brought to the mill "to learn the craft."[9] Did Vidalon's new hands learn to urinate in the vat to neutralize the acidity of the pulp, perhaps an apocryphal technique but a telling bit of know-how? Did the jeunes élèves, who needed more supervision, ever master their trade's shortcuts and fine points, just the sort of thing that the Montgolfiers' rationalized regime left out—or expunged? For papermaking remained a craft, a subtle mechanical art immune to Taylorist adulteration. It was also a capital-intensive industry. Skill was the fulcrum of both the art and the mill, and even Jean-Pierre cut corners in pursuit of it. He disparaged the journeyman Sauvade's overall mastery of his craft, reviled him as a man who "[did] not know how to occupy himself when he [was] not at the vat," but admitted that he was worth rehiring as a coucher in pressing circumstances.[10] In April 1780, Jean-Pierre asked Augustin if he intended to send Périgord to the big mill. If not, Jean-Pierre intended to hire a rente, the trade term for both the tramping man and the custom that nourished him.[11] Four years later, Jean-Pierre did engage Durif, a "*rente* passing by," as a temporary replacement.[12] Getting the work done sometimes got in the way of the new order.

The Montgolfiers would have settled for the manufacturer's utopia sketched by a humble Auvergnat papermaker in the first years of the nineteenth century:

> The workers would be more submissive, more exact, and more attentive to their tasks; the finish and fabrication of the merchandise would be far better; wages would diminish imperceptibly; the quota for the workday would be augmented to the liking of the master; the price of the merchandise would be more reasonable; the quantity of it would be more considerable; the manufacturer would do his best to improve; the government and society would find it to their own interest.[13]

Doubtless, Vidalon's *patrons* believed they had ventured as far in this direction as their reliance on skilled men permitted. So did Jubié, an Auvergnat inspector of manufactures. He welcomed the Montgolfiers' adoption of Hollander beaters as a step in the process that "could extinguish little by little the spirit of union which reigns among the workers of this province."[14]

PART FIVE

The End of Hand Papermaking

CHAPTER SEVENTEEN

The French Revolution and the Papermaking Machine

Revolutions run on paper. But in France, the craft's good times—produced by the rapid rise of political journalism, the general urge to comment, the outpourings of officials, and paper money itself—were quickly effaced by war and scarcity.[1] Inflation, lost markets, labor shortages, requisitioned mules, and above all, rare, expensive rags, left papermaking in turmoil. Even grandees like the Montgolfiers were not spared.

Michel Filliat, a petty producer from Antraigues, complained in December 1793 that he could not muster enough workers to continue making paper.[2] In response to a state survey, Vidalon's *patrons* carefully observed that all their workers "above twenty-five years of age [were] *pères de famille*."[3] Conscription claimed so many men, Michel-François Montgolfier lamented, that he had to idle one of the two vats at Vidalon-le-Bas.[4] And Revolutionary requisitions carried off more than men: François Johannot's workers grumbled in the summer of 1794 that the absence of mules caused "our lack of work."[5] When men and mules were available, food was not. Already in 1789, Vidalon's *patrons* had commuted the master's table into a cash payment, likely signaling the dearness of bread rather than an end to their paternalism.[6] His production halved, Michel-François Montgolfier finally located some hands, but for want of food, he could not hire them.[7] Scavenging for grain, an Auvergnat manufacturer moaned, cost him workers and output.[8]

Familiar markets were closed, or dangerous. Vidalon's *patrons* muttered about "the obstruction of markets," while unlucky Michel-François Montgolfier made sure to note that he had traded with "Nîmes, Montpellier and its environs, *Sans-Nom* [Marseilles], and *Ville-Affranchie* [Lyons] before the siege."[9] It was hardly worth pursuing the thin supply of old linen, especially as its price spiraled. In May 1792, a quintal of rags still cost the Montgolfiers 12 livres 4 sous. Over the next fifteen months, the price more than doubled to 27 livres 6 sous, doubtless compounding "the scarcity of primary materials." Rags then came under the *maximum* (state-set price ceilings), and in November 1793 the Montgolfiers paid an average of 11 livres 12 sous per quintal, or only 6 sous more than the rate established in their district. The overthrow of

fixed prices and the attendant inflation sent the cost of old linen skyward: 17 livres 16 sous per quintal in November 1794; 21 livres 2 sous in January 1795; 91 livres 4 sous in June; and a dizzying 731 livres 12 sous at the end of December. Dramatic deflation followed. By November 1796, the Montgolfiers had paid 8 livres 10 sous in reformed currency for a quintal of old linen. Thereafter, stability reigned. From 1797 through 1801, Vidalon's *patrons* bought rags during the months of May and November costing from 9 to 11 livres per hundredweight. By May 1803, a quintal of old linen cost the Montgolfiers 15 livres, but the same measure of discarded rags was available for 12 livres 10 sous six months later.[10] (The relatively benign price of old linen at the turn of the century was one secret of the regular productivity recorded in the livre de fabrication.)

The Revolution also revealed diverse commitments among the Montgolfiers themselves. Old Pierre remained a staunch royalist to the end, which came in 1793. Fearing the anti-clericalism that had risen in Annonay, he had moved the mill chapel to a small farm he owned, where he heard Mass conducted by a refractory priest.[11] For his part, Etienne continued to pursue perquisites from the state, most notably for counterfeit-proof assignats, the Revolution's paper money. He also turned his hand to the design of the subdivisions of the new department of the Ardèche, a task that nicely suited his rationalizing taste. Like his managerial approach, Etienne's politics bespoke measured, ordered change: he hovered on the cautious flank of the political center.[12] He died in 1799.

Jean-Pierre's fate had an odd, almost metaphorical twist. He labored tirelessly to keep the mill afloat during the ebbs and flows of the Revolution. Whatever politics he possessed were subsumed to commerce. In 1795, he headed to Voiron to collect his one-third share in the liquidation of his brother Joseph's mill. He took his compensation in assignats (paper!) and concealed them in the lining of his breeches. While taking shelter from a storm on his way home, he died. A month later, Joseph informed the family that his brother's pants had been rich in paper currency. By then, however, they had been laundered, and the certificates, reduced to pulp, were fit only for recycling in the Hollander beaters.[13]

Etienne had been a member of the National Guard, a Revolutionary institution which enlisted Vidalon's workers as well. Jean-Baptiste Montgolfier, Jean-Pierre's son, who had an increasingly active hand in managing the mill, held the rank of captain of the unit. Jean Couvat, the only worker who enjoyed the title of *maître-ouvrier* at Vidalon-le-Haut, was also the only lieutenant in the guard, suggesting a close parallel between the hierarchies. The ranks included sixteen of the twenty-nine male hands at work in the mill in early 1794,

with most of those not enrolled either older governors or adolescent apprentices.[14] Still, the workers' low standing in the guard did not necessarily translate into submissiveness in the shops. On October 23, 1798, one of Vidalon's *patrons* wrote, "All my workers have just quit me at the same time because I did not want to subscribe to the increase [in wages] that they wanted us to offer them." He reported that the workers at Vidalon-le-Bas had taken the same step, and urged his correspondent to send any journeymen his way who "wish to work."[15] Vidalon's hands soon returned to their tasks, evidently on the Montgolfiers' terms.[16] Their collective action did not draw on the "modes"; instead, it centered on the cash nexus that Vidalon's *patrons* trusted in the training of their new model workers.

The paperworkers of Revolutionary France continued to maintain their own republic. In 1794, the state, desperately in need of paper for its money and ordinances, decreed that "the coalitions among workers of different manufactories, in writing or by emissaries, in order to provoke the cessation of work, will be regarded as blows struck at the tranquility which must reign in the shops."[17] So much for fraternity. Like the Crown before them, France's Revolutionary regimes affirmed the papermakers' rights to train the apprentices and hire the journeymen they desired.[18] In 1796, in a resounding echo of the royal edict of 1777, the Directory denounced "all unlawful mobs composed of [paper]workers or incited by them against the free exercise of the industry or of work."[19] But the journeymen paperworkers continued their practices and held fast to their combination. Their republic, the steadfast "race" of journeymen and their ways, outlasted Louis XVI, Robespierre, and Napoleon. It was shredded by the machine.

Nicolas-Louis Robert patented the papermaking machine in 1799. His device consisted of continuous belts of woven wire mesh which picked up the pulp, shook it free of water, and deposited the infant sheet on a felt-covered roller. Simply put, Robert's machine mechanized movements that had been the monopoly of skilled men. While a deft vat crew took a full day to fashion five or six reams of paper, the output of the machine was measured in feet per minute. Improved versions of Robert's instrument commonly produced 25 to 40 feet of paper per minute in the 1840s, and as much as 150 in the 1870s.[20] Such productivity sealed the fate of a world of work that spanned thirteenth-century Fabriano and early nineteenth-century Vidalon-le-Haut: in 1867, 400 machines and only 140 vats furnished paper in France.[21] Yet, Robert had not created his machine solely, or even primarily, to increase output. According to Léger Didot, who employed Robert as a clerk in his mill at Essonnes, the device was a response to the paperworkers' esprit de corps. "Disgusted, like me, by the bad conduct of the *corporation* of paperworkers," Didot explained,

Robert decided "to seek the means of fabricating paper without their aid."[22] Whereas the Montgolfiers had designed their nouvel ordre to rid their shops of the workers' ways, Robert intended to rid the industry of the workers themselves.

Around 1820, Barthélemy Barou de Canson, Etienne Montgolfier's son-in-law and the principal proprietor of Vidalon-le-Haut, introduced his own new order at the mill. It, too, provoked the journeymen, but he was determined to liberate his shops from their "tyranny." Soon he fired all of Vidalon's veteran hands and engineered a work force composed of "strangers to the craft."[23] In 1822, he installed the first papermaking machine at Vidalon-le-Haut.[24] Evidently, however, he had not yet dispensed with the mill's vats and new skilled men. On May 31, 1823, the mill journal noted Canson's adherence to one trade custom, albeit a tradition defined as legitimate by the masters: "Gifts to the workers, cross of May."[25] He had honored Sainte-Croix, the craft's day, the day when the bosses recognized their obligation to their skilled hands. But the era of the skilled paperworker was passing. No longer distinguished by their skills, they affiliated with a brotherhood of craftsmen from diverse trades, the *compagnons du devoir*.[26]

Conclusion

The tramping paperworker, the Montgolfiers knew, was a masterless man. He moved about freely, regardless of whether he carried a state-mandated discharge from his previous employer. His arrival at a mill set off drinking sprees and reminded the domiciled workers of their obligations to distant brothers. He traded on skills inherited from his father and rituals handed down along dusty roads and in dark taverns, beyond the masters' vision.

The Montgolfiers longed for a new model worker—the employee. He would toil as they wished, with the skills they taught him. Immune to the appeals of his fellows, he would recognize that his interests and those of his boss were one. He would sweat over makeshift work, awaiting rejuvenating rains with his master. He was the most original feature of the Montgolfiers' nouvel ordre.

Even innovative programs of labor discipline have deep roots in the past. Whether prompted by mechanization or market pressures, there are old practices and values to discard as well as new skills and habits to absorb. Abstract principles mingle with memories of pranks, confrontations, and rushed work. Touched by the humorless, machinelike workshops depicted in the *Encyclopédie*, Vidalon's *patrons* tried to recreate them in their mill. Rather than lore, the empirical wisdom of skilled hands, their ateliers would be governed by law, experimentally verified techniques. But their governors continued to judge the readiness of the pulp by its resemblance to the hairs on a fly's leg and their sizerman studied the impression left by his tongue on newly finished sheets.

For men so committed to a dispassionate approach to their art, the Montgolfiers nevertheless steeped their new regime in their own passions and experiences. It rested on Pierre's firm paternalism, catechism classes, and delight in improved techniques, especially the mechanical. It drew on Etienne's inclinations toward calculated reform, intellectual companionship with Desmarest, and penchant for perquisites and the corridors of power. And it turned on Jean-Pierre's tour de France, canny appraisals of workers, and tolerance of some shopfloor turbulence—so long as it stayed within bounds.

The Montgolfiers pursued every opportunity to expand their dominion

over Vidalon's shops and commerce. They alone would determine techniques and tools, the weights and dimensions of their reams, and the assignment of tasks and stations. They equipped the mill with Hollander beaters, tables and presses for échange, home-grown hands and hand-picked journeymen, and a flock of female workers to shuffle their Dutch-style reams. They trusted in advanced incentives, a carefully partitioned workday, and precise mandates and penalties. They outlawed the festive "modes" and fines levied by journeymen, and had apprentices and veterans alike vow to distance themselves from the masters' affairs and those of their fellows. They improvised, selectively adopted state directives, even turned to the "licit" customary heritage of the craft. Their nouvel ordre was no less novel for all that. Rather than a stage in an artificial lineage of labor discipline, it was an attempt to restore and invent the proper order of their trade. Mirroring its architects, Vidalon's new order was ambitious, austere, and authoritarian, yet characterized by the timely advance as well as the cold shoulder. Above all, the Montgolfiers' impersonal disciplinary codes and paternalism would undercut the journeymen's "singular species of police," that "perpetual vexation so prejudicial to the progress of the *fabriques*."[1]

The state seconded and sustained certain of the Montgolfiers' efforts. Its system of preferments and subventions removed much of the risk in daring. It recognized the utility of the rational play of private interest in papermaking, effectively deregulating both markets and technology. It provided the technical promoter and guide, Desmarest and Ecrevisse, who helped Vidalon's *patrons* obtain their subsidy and get the thump of the Hollander beaters right. Together, the state and applied science empowered the Montgolfiers and amplified their confidence. If Versailles had removed internal tolls and tariffs, and done more than issue paper mandates to rein in the journeymen, the Montgolfiers would have had the larger regime they desired.

Skill and its culture, however, remained in their path. Consequently, when the walkout of 1781 provided Vidalon's *patrons* with the chance to shake free of the workers' ritual and association, they seized it eagerly. While they remodeled their shops, they recruited newcomers, innocent of the "modes" and time-worn techniques. They replaced a worker-dominated labor market, organized to keep the ranks thin, familial, and initiated, with an internal labor market geared to serve their ends. They schooled the greenhorns in steady intensity and their version of the art. This was neither deskilling nor Taylorism; it was the fitting of new men with new skills, new wants, and perhaps new loyalties. Shorn of their cultural patrimony, the Montgolfiers' skilled, wage-earning hands would truly become proletarians.

The Montgolfiers' latitude for technological innovation was as wide as

their dependence on skilled workers allowed. They had secured a productive, persistent work force of relatively mild-mannered men. But they were men, and Vidalon's shops still echoed with their tumult and the cabarets of Annonay still beckoned. They sweated, sang, and died within the earshot of Adélaïde de Montgolfier, but beyond her hearing. If her family shared Wedgwood's dream of making "machines of men," they had not succeeded. That was one reason Canson installed the machine itself. The dream remains an illusion.

APPENDIX

Tables and Graph

TABLES

GRAPH

Note: In all tables except Table H, all numbers are rounded to the nearest tenth.

Table A
Annual Cost of Operating a Working Vat according to Lalande

Item	Cost in livres tournois	Percentage of Total Cost
Labor		
Male workers (food and salary)	1,365	18.8%
Female workers	463	6.4
Furnishings and raw materials		
Rags	4,800	65.9
Size	210	2.9
Alum	40	0.5
Felts	150	2.1
Wood, Coal	150	2.1
Upkeep of the Mill, Grease, Soap	100	1.4

Source: Lalande, *Art of Papermaking,* 59–60.

Table B
Cost of Upkeep at the Terrace Vat for July 1780

Item	Cost in livres tournois	Percentage of Total Cost
Labor		
Food for 1 governor and 3 workmen	66 livres	9.0%
Food for 1 apprentice	10 livres 10 sous	1.4
Base pay for 1 governor and 3 workmen	25 livres 10 sous	3.5
Odd jobs	2 livres	0.3
Incentive payment to the team (probably including the governor)	27 livres 4 sous 4 deniers	3.7
Payment to the governor for work on a Sunday night	1 livre 10 sous	0.2
Sizing room women	25 livres	3.4
Furnishings and raw materials		
Rags	543 livres 16 sous	74.3
Coal	20 livres	2.7
Candles	6 livres 16 sous	0.9
Grease	2 livres 6 sous	0.3
Soap	16 sous	0.1
Felts	13 sous	0.1

Source: AN, 131 MI 53 AQ 92, JG, July 31, 1780.

Table C
Age Distribution of the New Men at Entry, 1781–1785

Age in Years	Number of Men (*N* = 90)	Percentage
0–14	7	7.8%
15–19	40	44.4
20–24	19	21.1
25–29	11	12.2
30–34	4	4.4
35–39	3	3.3
40–44	2	2.2
45 and over	1	1.1
Unknown	3	3.3

Source: AN, 131 MI 53 AQ 23, document 14.

Table D
Stability of the New Men, April 1784–December 1789

Months of Work	Number of Workers (*N* = 34)	Percentage
1–12	2	5.9%
13–24	2	5.9
25–36	7	20.6
37–48	4	11.8
49–60	4	11.8
61–72	4	11.8
73–84	4	11.8
85–96	4	11.8
More than 96	3	8.8

Source: AN, 131 MI 53 AQ 24, JO.
Note: Chirol's son, who shared his father's number in the Montgolfiers' accounts, cannot be traced.

Table E
Tramping Time Prior to Employment at Vidalon-le-Haut, April 1784–December 1789

Days per Journey	Number of Workers (*N* = 33)	Percentage
0–4	9	27.3%
5–9	5	15.1
10–14	6	18.2
15–19	4	12.1
20–24	3	9.1
33	1	3.0
53	1	3.0
65	1	3.0
90	1	3.0
115	1	3.0
124	1	3.0

Source: AN, 131 MI 53 AQ 24, JO.

Table F
Stability of Workers Hired with a Certificat de Congé, April 1784–December 1789

Months of Work	Number of Workers (N = 54)	Percentage
0–3	28	51.9%
4–6	8	14.9
7–12	8	14.9
13–24	3	5.5
25–36	3	5.5
37–48	2	3.7
49 or more	2	3.7

Source: AN, 131 MI 53 AQ 24, JO.

Table G
The Montgolfiers' Evaluations of Male Paperworkers and Governors, April 1784–December 1789

Evaluation	Number of Workers
Excellent work/worker	2
Good work/worker or performed his task well	8
Good enough work/worker or performed his task well enough	8
Passable work/worker	5
Ordinary or mediocre work/worker	6
Bad work/worker	3
Lazy	7

Source: AN, 131 MI 53 AQ 24, JO.

Note: Vidalon's *patrons* described three men as *adroit;* one of these men was also praised for doing his work well, but another was condemned as lazy.

Two of the men described by the Montgolfiers as lazy were also reproached as mediocre workers; a third was portrayed as a bad worker.

Counted among the bad workers, Thibert inched up to the rank of mediocre worker when he was watched.

Table H
The Work Year: Aggregate Days Worked per Year per Team

Year	Team	Days Worked	Breach Days
1799	1	268	33
	2	267	29
	3	254	43
	4	259	38
	5	266	28
1800	1	241	36
	2	237	40
	3	236	42
	4	241	38
	5	237	37
1801	1	265	37
	2	263	35
	3	266	38
	4	262	41
	5	230	71
	Extra (2/16–4/23 only)	36	16
1802	1	235	42
	2	232	33
	3	226	49
	4	230	52
	5	219	70
1803	1	224	37
	2	219	31
	3	215	34
	4	205	46
	5	187	50
1804	1	271	31
	2	264	39
	3	246	46
	4	259	37
	5	238	66
	Extra (4/16–12/31)	147	55
1805	1	248	12
	2	254	14
	3	253	11
	4	246	12
	5	221	37
	Extra (1/1–6/18)	91	10

Source: AN, 131 MI 53 AQ 265 and 266.

Note: Since the Montgolfiers organized production by teams, this table assesses their work. Accordingly, when a freeze or scarce water led Vidalon's *patrons* to shift the second team to the fifth production vat, the workday was credited to the second team.

All numbers are rounded to the nearest whole.

Table I
Holiday Calendar at Vidalon-le-Haut, 1799–1805

Holiday	1799	1800	1801	1802	1803	1804	1805
New Year's Day	X	X	X	X	–	–	–
Epiphany	O	X	X	X	–	–	–
Purification Day	X	O	X	X	–	–	–
Shrove Monday	–	–	–	–	–	–	–
Shrove Thursday	–	–	–	–	–	–	–
Ash Wednesday	#	#	#	#	#	#	#
Annunciation	X	X	X	X	–	–	–
Good Friday	#	#	#	#	#	#	#
Easter Monday	X	X	X	X	X	X	X
Easter Tuesday	–	X	X	X	–	–	–
Ascension	X	X	X	X	X	X	X
Pentecost	X	X	X	X	X	X	X
Sainte-Croix	X	X	X	X	–	–	–
Feast of Saint John	X	X	X	X	–	–	–
Feast of Peter and Paul	X	O	X	X	–	–	–
Assumption	X	X	X	O	X	X	X
Nativity of the Virgin (September 8)	X	*	X	X	–	–	–
All Saints' Day	X	X	O	X	X	X	X
All Souls' Day	#	#	#	#	#	#	#
Immaculate Conception	O	X	X	X	–	–	*
Christmas	X	X	X	X	O	X	*
December 26	X	X	X	O	X	X	*
December 27	X	X	O	–	–	–	*

Source: AN, 131 MI 53 AQ 265 and 266.

Key: X = Celebrated with a whole day off.
O = Probably celebrated but holiday fell on a Sunday.
– = Did not celebrate.
= Partial or complete workday combined with distinctive meals and pay rate.
* = Data are missing.

Table J
Monthly Rhythm of Production at Vidalon-le-Haut, 1799–1805

Month	Average Days Worked
January	22.0
February	21.5
March	24.1
April	23.1
May	22.9
June	18.5
July	17.8
August	18.0
September	14.1
October	19.5
November	22.1
December	22.4

Source: AN, 131 MI 53 AQ 265 and 266.

Note: For the purpose of statistical consistency, I have not included the work of the sixth team in this table. The Montgolfiers assembled this crew when demand exceeded capacity of the five regular units. A sixth team toiled in January 1805; February 1801 and 1805; March 1801 and 1805; April 1801, 1803, and 1804; May 1804 and 1805; June 1804 and 1805; July 1804; August 1804; September 1804, October 1804; November 1804; and December 1804.

August and September figures do not include 1800, and the December figure does not include 1805; information for these months was incomplete.

Table K
Quota-making at Vidalon-le-Haut by Day of the Week, 1803 and 1804

Days	Monday	Tuesday	Wednesday	Thursday	Friday	Saturday	Total
				1803			
Over quota	3	7	8	10	6	5	39
	(1.7%)	(4.2)	(4.7)	(5.9)	(3.5)	(2.8)	(3.8)
At quota	79	104	85	91	104	96	559
	(45.7)	(62.7)	(49.4)	(53.8)	(61.9)	(53.0)	(54.3)
One post under quota	69	40	56	55	49	69	338
	(39.9)	(24.1)	(32.6)	(32.5)	(29.2)	(38.1)	(32.8)
More than one post under quota	20	12	18	9	6	7	72
	(11.6)	(7.2)	(10.5)	(5.3)	(3.6)	(3.9)	(7.0)
Unable to determine	2	3	5	4	3	4	21
	(1.2)	(1.8)	(2.9)	(2.4)	(1.8)	(2.2)	(2.0)
Total complete days	173	166	172	169	168	181	1,029
				1804			
Over quota	5	8	14	11	8	8	54
	(2.0)	(3.4)	(6.2)	(4.7)	(3.5)	(3.3)	(3.8)
At quota	77	171	123	97	141	147	756
	(31.4)	(71.8)	(54.6)	(41.8)	(62.3)	(61.0)	(53.7)
One post under quota	145	40	61	101	56	67	470
	(59.2)	(16.8)	(27.1)	(43.5)	(24.7)	(27.8)	(33.4)
More than one post under quota	18	19	27	23	21	19	127
	(7.3)	(8.0)	(12.0)	(9.9)	(9.3)	(7.9)	(9.0)
Total complete days	245	238	225	232	226	241	1,407

Source: AN, 131 MI 53 AQ 265 and 266.

Note: The quotas appear in the *tarifs* issued by the Montgolfiers in October and November 1789 (AN, 131 MI 53 AQ 24, JO), and in AN, 131 MI 53 AQ 23, document 15. Vidalon's masters frequently modified these quotas. In doing so, they took account of the weight and tint of the paper, the quality and nature of the raw material, and the manner in which the rags were pulped; inevitably, however, they failed to record their reduced or enhanced expectations in the output register. Thus, if a vat crew turned out a particular sort of paper at an unvarying pace for several weeks, and if this rate deviated little from the posted goals and provoked no complaint from the masters, I have assumed that this new pace was the one the masters wanted—or accepted.

Table L
Number of Posts Produced of Each Quality of Paper by Vat in 1799

	Quality of Paper				
Vat	*Surfin*	*Fin*	*Moyen*	*Bulle*	*Trasse*
1	2,033	1,526	1,444	160	
	453	0	0	0	
2	216	436	0		
	633	2,251	292		
3	0	0	0		
	2,465	2,088	649		
4	455	0	327		
	1,748	2,729	462		
5			135	0	
			1,393	1,455	
6		0	531	0	0
		65	3,224	206	585

Source: AN, 131 MI 53 AQ 265 and 266.
Note: Where two numbers appear, the top number represents the number of posts that will be turned into finished reams weighing 40 pounds or more. A blank space indicates no production of that type of paper by the vat.

Production by Day of the Week at Vidalon-Le-Haut, 1799–1805

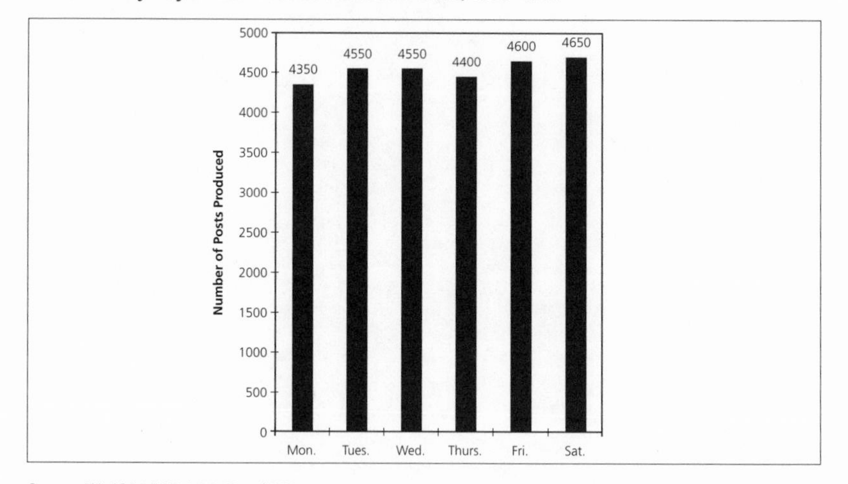

Source: AN, 131 MI 53 AQ 265 and 266.

Notes

PREFACE

1. Cited in Charles Gillispie, *The Montgolfier Brothers and the Invention of Aviation, 1783–1784* (Princeton: Princeton University Press, 1983), 8.
2. On this turmoil, see Charles-Moïse Briquet, "Associations et grèves des ouvriers papetiers en France aux XVIIe et XVIIIe siècles," *Revue internationale de sociologie* 5 (March 1897): 161–90; Henri Gachet, "Les Grèves d'ouvriers papetiers en France au XVIIIème siècle jusqu'à la Révolution," in *Papers of the Twelfth International Congress of the International Association of Paper Historians* (Haarlem, 1972), 125–48; and Henri Gazel, *Les Anciens Ouvriers papetiers d'Auvergne* (Clermont-Ferrand: A. Dumont, 1910).
3. On the question of early systems of labor discipline, see Michelle Perrot, "The Three Ages of Industrial Discipline in Nineteenth-Century France," in *Consciousness and Class Experience in Nineteenth-Century Europe*, ed. John Merriman (New York: Holmes and Meier, 1979), 149–68; Sidney Pollard, *The Genesis of Modern Management: A Study of the Industrial Revolution in Great Britain* (Cambridge: Harvard University Press, 1965); and E. P. Thompson, "Time, Work-Discipline, and Industrial Capitalism," in his *Customs in Common* (New York: New Press, 1991), 352–403. For more detailed local studies, see Pierre Caspard, *La Fabrique-Neuve de Cortaillod: Entreprise et profit pendant la Révolution industrielle, 1752–1854* (Paris: Publications de la Sorbonne, 1979); Serge Chassagne, *Oberkampf: Un Entrepreneur capitaliste au siècle des Lumières* (Paris: Aubier-Montaigne, 1980); Maurice Hamon and Dominique Perrin, *Au Coeur du XVIIIe siècle industriel: Condition ouvrière et tradition villageoise à Saint-Gobain* (Paris: Editions P.A.U., 1993); and Joan Scott, *The Glassworkers of Carmaux: French Craftsmen and Political Action in a Nineteenth-Century City* (Cambridge: Harvard University Press, 1974).
4. Alain Cottereau, "The Distinctiveness of Working-Class Cultures in France," in *Working-Class Formation: Nineteenth-Century Patterns in Western Europe and the United States*, ed. Ira Katznelson and Aristide Zolberg (Princeton: Princeton University Press, 1986), 111–54.
5. Thompson, *Customs in Common*, 386.
6. For the distinctive features of hand papermaking, see Donald Coleman, *The British Paper Industry, 1495–1860: A Study in Industrial Growth* (Oxford: Clarendon Press, 1958); Lucien Febvre and Henri-Jean Martin, *L'Apparition du livre*

(Paris: Albin Michel, 1958); Dard Hunter, *Papermaking: The History and Technique of an Ancient Craft*, 2d ed. (1947; rpr. New York: Dover Publications, 1978); and Judith McGaw, *Most Wonderful Machine: Mechanization and Social Change in Berkshire Paper Making, 1801–1885* (Princeton: Princeton University Press, 1987).

7. On the representation of work in the *Encyclopédie*, see William Sewell Jr., "Visions of Labor: Illustrations of the Mechanical Arts before, in, and after Diderot's *Encyclopédie*," in *Work in France: Representations, Meaning, Organization, and Practice*, ed. Steven Kaplan and Cynthia Koepp (Ithaca: Cornell University Press, 1986), 258–86.

CHAPTER 1: FRENCH INDUSTRY IN THE EIGHTEENTH CENTURY

1. On the Colbertian legacy and eighteenth-century French industry, see Jean-Pierre Hirsch, *Les Deux Rêves du Commerce: Entreprise et institution dans la région lilloise, 1780–1860* (Paris: Editions de l'Ecole des hautes études en sciences sociales, 1991); and Philippe Minard, *La Fortune du colbertisme: Etat et industrie dans la France des Lumières* (Paris: Fayard, 1998). Also of value is Jean-Louis Bourgeon, "Colbert et les corporations: L'Exemple de Paris," in *Un Nouveau Colbert*, ed. Roland Mousnier (Paris: SEDES/CDU, 1985).

2. Peter Jones, *Reform and Revolution in France: The Politics of Transition, 1774–1791* (Cambridge: Cambridge University Press, 1995), 1–11. The transformation of French institutions in general and of its craft and industrial production during the twilight of the Old Regime has generated an enormous bibliography. Among the key titles are Ken Alder, *Engineering the Revolution: Arms and Enlightenment in France, 1763–1815* (Princeton: Princeton University Press, 1997); P. W. Bamford, *Privilege and Profit: A Business Family in Eighteenth-Century France* (Philadelphia: University of Pennsylvania Press, 1988); Gail Bossenga, *The Politics of Privilege: Old Regime and Revolution in Lille* (Cambridge: Cambridge University Press, 1991); François Crouzet, *Britain Ascendant: Comparative Studies in Franco-British Economic History* (Cambridge: Cambridge University Press, 1990); Pierre Deyon and Philippe Guignet, "The Royal Manufactures and Economic and Technological Progress in France before the Industrial Revolution," *Journal of European Economic History* 9 (Winter 1980): 611–32; Philippe Guignet, *Mines, manufactures et ouvriers du Valenciennois au XVIIIe siècle* (New York: Arno Press, 1977); John Harris, *Essays in Industry and Technology in the Eighteenth Century: England and France* (Brookfield, Vt.: Variorum, 1992); Liliane Hilaire-Pérez, "Invention and the State in Eighteenth-Century France," *Technology and Culture* 32 (October 1991): 911–31; Steven Kaplan, "Social Classification and Representation in the Corporate World of Eighteenth-Century France: Turgot's 'Carnival,'" in *Work in France*, 176–228; Gwynne Lewis, *The Advent of Modern Capitalism in France, 1770–1840: The Contribution of Pierre-François Tubeuf* (Oxford: Clarendon Press, 1993); Harold Parker, *An Administrative Bureau during the Old Regime* (Newark: University of Delaware Press, 1993); and Pierre Renouvin, *Les Assemblées provinciales de 1787: Origines, développement, résultats* (Paris: A. Picard, 1921).

3. Quoted in Steven Kaplan, "The Luxury Guilds in Paris in the Eighteenth Century," *Francia* 9 (1981): 288.

4. T. C. W. Blanning, *The French Revolution: Aristocrats versus Bourgeois?* (London: Macmillan, 1987), 11.

5. Pierre Léon, "La Réponse de l'industrie," in *Histoire économique et sociale de la France, 1660–1789*, 2, ed. Fernand Braudel and Ernest Labrousse (Paris: Presses universitaires de France, 1970), 261.

6. Archives départementales de l'Ardèche (henceforth ADA), C 143, Antoine-François Montgolfier to the Estates of Languedoc, 1780.

7. David Landes, *The Unbound Prometheus: Technological Change and Industrial Development in Western Europe from 1750 to the Present* (Cambridge: Cambridge University Press, 1969), 132.

8. This description is drawn, in part, from William Sewell Jr., "Artisans, Factory Workers, and the Formation of the French Working Class, 1789–1848," in *Working-Class Formation,* 46–50.

9. Léon, "La Réponse de l'industrie," 227.

10. Colin Jones, "Bourgeois Revolution Revivified: 1789 and Social Change," in *Rewriting the French Revolution,* ed. Colin Lucas (Oxford: Clarendon Press, 1991), 88–93; and Cissie Fairchilds, "The Production and Marketing of Populuxe Goods in Eighteenth-Century Paris," in *Consumption and the World of Goods,* ed. John Brewer and Roy Porter (London: Routledge, 1993), 228–48.

11. On liberal ideology and action, see Keith Baker, *Condorcet: From Natural Philosophy to Social Mathematics* (Chicago: University of Chicago Press, 1975); Douglas Dakin, *Turgot and the Ancien Régime in France* (London: Methuen, 1939); Elizabeth Fox-Genovese, *The Origins of Physiocracy: Economic Revolution and Social Order in Eighteenth-Century France* (Ithaca: Cornell University Press, 1976); Simone Meyssonier, *La Balance et l'horloge: La Genèse de la pensée libérale en France au XVIIIe siècle* (Montreuil: Editions de la Passion, 1989); and William Sewell Jr., *Work and Revolution in France: The Language of Labor from the Old Regime to 1848* (Cambridge: Cambridge University Press, 1980), 62–91. For the critique of the craft communities and the guilds' response, see Kaplan, "Social Classification," passim.

12. Harris, *Essays in Industry and Technology*; and Léon, "La Réponse de l'industrie," 239–46.

13. Michael Sonenscher, *Work and Wages: Natural Law, Politics, and the Eighteenth-Century French Trades* (Cambridge: Cambridge University Press, 1989), 99–209.

14. Sean Wilentz, *Chants Democratic: New York City and the Rise of the American Working Class, 1788–1850* (New York: Oxford University Press, 1984), 12.

15. Recent scholarship has been preoccupied with the place of cottage industry in the lineage of mass production. This trend has fed the relative neglect of consolidated manufacture before full mechanization. For a similar view, see the introduction to *Manufacture in Town and Country before the Factory*, ed. Maxine Berg, Pat Hudson, and Michael Sonenscher (Cambridge: Cambridge University Press, 1983), 19–20.

16. On these ties, see, above all, Charles Gillispie, *Science and Polity in France at the End of the Old Regime* (Princeton: Princeton University Press, 1980), 388–478.
17. Perrot, "Three Ages of Industrial Discipline," 150–56.
18. Maurice Dobb, *Studies in the Development of Capitalism* (London: George Routledge and Sons, 1946), offers a mechanical depiction of such transitions. Instead, my approach to these transformations is heavily indebted to Sewell, "Visions of Labor," esp. 275–78, and the complex connections he draws between changes in technology, shifts in the organization of production, and proletarianization. For more extensive discussions of proletarianization, see chs. 5, 7–11.

CHAPTER 2: MAKING PAPER

1. Sonenscher, *Work and Wages,* 186–87, 205–9.
2. Joseph-Jérôme Lefrançois de Lalande, *The Art of Papermaking*, trans. Richard Atkinson (Kilmurry, Ireland: Ashling Press, 1976), 56. Lalande originally published his *Art de faire le papier* in 1761. I have modernized the capitalization in Atkinson's translations.
3. Archives nationales (henceforth AN), 131 MI(crofilm) 53 AQ 30.
4. AN, 131 MI 53 AQ 23, document 7.
5. Lalande, *Art of Papermaking*, 5–6.
6. AN, 131 MI 53 AQ 89, "Journal général" (henceforth JG), May 23, 1761.
7. Lalande, *Art of Papermaking*, 8.
8. AN, 131 MI 53 AQ 23, document 7.
9. Lalande, *Art of Papermaking*, 18.
10. Lalande, *Art of Papermaking*, 26.
11. AN, 131 MI 53 AQ 23, document 9.
12. AN, 131 MI 53 AQ 23, document 7.
13. For the costs of the molds used at Vidalon-le-Haut, see AN, 131 MI 53 AQ 23, document 23.
14. On the French state's regulation of papermaking, see ch. 4.
15. Nicolas Desmarest, "Papier (Art de fabriquer le)," in *Encyclopédie méthodique: Arts et métiers mécaniques*, 5 (Paris, 1788), 550.
16. On the Montgolfiers' willingness to place their customers' needs ahead of the state's mandates, see ch. 4.
17. For the trade's colorful vocabulary, see Lalande, *Art of Papermaking*, passim; E. J. Labarre, *Dictionary and Encyclopaedia of Paper and Paper-making*, rev. 2d ed. (London: Oxford University Press, 1952); and J.-L. Boithias and C. Mondin, *Les Moulins à papier et les anciens papetiers d'Auvergne* (Nonette: Editions Créer, 1981).
18. Desmarest, "Papier," 511.
19. Lalande, *Art of Papermaking*, 39.
20. For the paperworkers' ailments, see McGaw, *Most Wonderful Machine*, 50, 53.
21. Desmarest, "Papier," 511.

22. Lalande, *Art of Papermaking*, 41.
23. AN, 131 MI 53 AQ 23, document 7.
24. Lalande, *Art of Papermaking*, 43.
25. Lalande, *Art of Papermaking*, 46.
26. Lalande, *Art of Papermaking*, 45.
27. Lalande, *Art of Papermaking*, 46.
28. AN, 131 MI 53 AQ 23, document 17.
29. AN, 131 MI 53 AQ 23, document 7.
30. Lalande, *Art of Papermaking*, 46.
31. AN, 131 MI 53 AQ 23, document 7.
32. Lalande, *Art of Papermaking*, 50.
33. Desmarest, "Papier," 524.
34. For the general characteristics of each grade, see Lalande, *Art of Papermaking*, 54.
35. Robert Darnton, *The Kiss of Lamourette: Reflections in Cultural History* (New York: Norton, 1990), 140–41.
36. Lalande, *Art of Papermaking*, 56.
37. Desmarest, "Papier," 522.
38. Sewell, *Work and Revolution*, 64–72, provides a brief, admirable discussion of the mechanical arts in the Enlightenment, largely from Diderot's perspective.
39. Harry Braverman, *Labor and Monopoly Capital: The Degradation of Work in the Twentieth Century* (New York: Monthly Review Press, 1974), 70–152.

CHAPTER 3: THE MONTGOLFIERS AND THEIR CRAFT

1. Febvre and Martin, *L'Apparition du livre*, 465.
2. The bibliography on innovative systems of labor discipline is enormous. Valuable entry points include "Naissance de la classe ouvrière en France," special issue of *Le Mouvement social* 97 (October–December 1976); Chassagne, *Oberkampf;* Neil McKendrick, "Josiah Wedgwood and Factory Discipline," *Historical Journal* 4, no. 1 (1961): 30–55; Pollard, *Genesis of Modern Management*, esp. ch. 5; and Rolande Trempé, *Les Mineurs de Carmaux, 1848–1914* (Paris: Editions Ouvrières, 1971).
3. AN, F^{12} 1477, "Mémoire du sieur Montgolfier fabricant et propriétaire d'une fabrique à Annonay."
4. AN, 131 MI 53 AQ 22, document 6.
5. Landes, *Unbound Prometheus*, 131–33.
6. AN, 131 MI 53 AQ 23, document 9.
7. Quoted in Charles Tilly, *The Contentious French: Four Centuries of Popular Struggle* (Cambridge: Harvard University Press, 1986), 192.
8. Gillispie, *Montgolfier Brothers*, 5, 7.
9. Tilly, *Contentious French*, 194.

10. ADA, C 960, Nicolas Desmarest, "Mémoire sur l'état actuel des papeteries d'Annonay et sur les améliorations dont elles sont susceptibles."
11. Archives départementales de l'Hérault (henceforth ADH), C 2676, "Etat des moulins à papier."
12. Desmarest, "Mémoire sur l'état actuel."
13. "Etat des moulins à papier."
14. AN, 131 MI 53 AQ 28, offers numerous examples of the Montgolfiers' willingness to share information with their confrères. A few notable cases are the letter to Grand on June 19, 1786; to Thollet on October 20, 1786; and to an Ambertois manufacturer on October 24, 1786.
15. AN, 131 MI 53 AQ 27, Etienne Montgolfier to Jovin of Saint-Etienne, August 18, 1785.
16. AN, 131 MI 53 AQ 27, letters of June 27, 1785, to Joubert and Palhion.
17. AN, 131 MI 53 AQ 22, document 6.
18. For the history of the Montgolfier family, see Léon Rostaing, *La Famille de Montgolfier, ses alliances, ses descendants*, 2d rev. ed. (Lyons: Presse Lyonnaise du Sud-Est, 1933). On the marriage, see Marie-Hélène Reynaud, *Les Moulins à papier d'Annonay à l'ère pré-industrielle, les Montgolfier et Vidalon* (Annonay: Editions du Vivarais, 1981), 21.
19. Reynaud, *Moulins à papier*, 23.
20. On the two Vidalons, see Reynaud, *Moulins à papier*; on the small scale of most French paper mills in 1812, see Louis André, *Machines à papier: Innovation et transformations de l'industrie papetière en France, 1798–1860* (Paris: Editions de l'Ecole des hautes études en sciences sociales, 1996), 31.
21. Reynaud, *Moulins à papier*, 28.
22. Rostaing, *La Famille de Montgolfier*, 71.
23. Rostaing, *La Famille de Montgolfier*, 74–75.
24. AN, 131 MI 53 AQ 22, document 6, is one of many sources that reveal Pierre Montgolfier's embrace of applied science as well as technological innovation.
25. Reynaud, *Moulins à papier*, 29–40.
26. Reynaud, *Moulins à papier*, 38.
27. Bibliothèque de l'Arsenal, ms. 5947, fol. 50, Emeric David, "Mon Voyage de 1787."
28. Family archives of Régis de Montgolfier (courtesy of Charles Gillispie), letter of August 1786.
29. Gillispie, *Montgolfier Brothers*, 39.
30. Reynaud, *Moulins à papier*, 35–38.
31. Reynaud, *Moulins à papier*, 59–60.
32. Pollard, *Genesis of Modern Management*, 12–13, 23, 123.
33. AN, 131 MI 53 AQ 23, document 10.
34. "Etat des moulins à papier."
35. Jean-Louis Flandrin, *Families in Former Times*, trans. Richard Southern (Cambridge: Cambridge University Press, 1979), 87.

36. Reynaud, *Moulins à papier*, 137.
37. Reynaud, *Moulins à papier*, 62.
38. Jones, "Bourgeois Revolution Revivified," 109.
39. AN, 131 MI 53 AQ 22, document 5.
40. AN, F^{12} 1474, "Mémoire sur les papeteries du Dauphiné et sur l'insubordination des ouvriers," 1771 (?).
41. "Etat des moulins à papier."
42. "Etat des moulins à papier."
43. "Mémoire du sieur Montgolfier fabricant et propriétaire d'une fabrique à Annonay."
44. ADH, C 2671, "Etat des moulins ou manufactures à papier situés sur la rivière de Deûme, lès Annonay."
45. Robert Darnton, *The Business of Enlightenment: A Publishing History of the Encyclopédie, 1775–1800* (Cambridge: Harvard University Press, 1979), 187.
46. Lalande, *Art of Papermaking*, 56.
47. These figures were drawn from AN, 131 MI 53 AQ 265 and 266.
48. Reynaud, *Moulins à papier*, 194, Table.
49. AN, 131 MI 53 AQ 27.
50. "Etat des moulins à papier."
51. ADA, C 143, letter of January 7, 1780.
52. ADH, C 2679, "Mémoire sur l'état actuel des papeteries de MM. Montgolfier et Johannot."
53. Sonenscher, *Work and Wages*, 213.

CHAPTER 4: RAGS, REGULATION, AND GOVERNMENT STIMULATION

1. ADA, C 143, letter of January 7, 1780.
2. On the complex ties between manufacturers and the eighteenth-century French state, see Deyon and Guignet, "The Royal Manufactures," passim; Gillispie, *Science and Polity*, 388–478; Hilaire-Pérez, "Invention and the State," passim; P. Jones, *Reform and Revolution*, 88–96; Minard, *La Fortune du colbertisme*; idem, "La Fin de l'inspection des manufactures: Premières hypothèses sur le dérèglement d'une institution du commerce," in *Etat, finances et économie pendant la Révolution française* (Paris: Comité pour l'histoire économique et financière de la France, 1991), 295–303; idem, "L'Inspection des manufactures et la réglementation industrielle à la fin du XVIIIe siècle," in *Naissance des libertés économiques*, ed. Alain Plessis (Paris: Institut d' histoire de l'industrie, 1993), 49–60; Parker, *An Administrative Bureau*; idem, *The Bureau of Commerce in 1781 and Its Policies with Respect to French Industry* (Durham, N.C.: Carolina Academic Press, 1979); and Sonenscher, *Work and Wages*, 210–18.
3. A. Latour, "Paper: A Historical Outline," *Ciba Review* 72 (1949): 2635. A comprehensive history of hand papermaking in Europe remains to be written. Meanwhile, useful starting points are Febvre and Martin, *L'Apparition du livre*, and Hunter, *Papermaking*.

4. Febvre and Martin, *L'Apparition du livre*, 43.
5. Coleman, *British Paper Industry*, 65.
6. Coleman, *British Paper Industry*, 22, esp. n. 2.
7. "Etat des moulins à papier de l'Auvergne en 1717," cited in Boithias and Mondin, *Les Moulins à papier . . . d'Auvergne*, 56.
8. Quoted in Coleman, *British Paper Industry*, 22, n. 3.
9. Jean-Baptiste Colbert, in *Lettres, instructions, et mémoires de Colbert,* ed. Pierre Clément (Paris: Imprimerie nationale, 1861–73). See Colbert's "Mémoire sur le commerce," (1664), 2: cclxix; and "Mémoire au Roi sur les finances," (1670), 7: 240.
10. Febvre and Martin, *L'Apparition du livre*, 48.
11. "Etat des moulins à papier."
12. Pierre Claude Reynard, "*Intendants, Subdélégués, et Fabricants de Papier*: Hesitant Dialogues—Administrative Practices in the Making," unpublished manuscript, 19.
13. Coleman, *British Paper Industry*, 169, Table.
14. The number of rag suppliers in 1789 was compiled from AN, 131 MI 53 AQ 95, "Journal général" (henceforth JG), 1789. The rate of increase in the price paid by the Montgolfiers for discarded, superfine linen can be found in Reynaud, *Moulins à papier*, 243.
15. AN, 131 MI 53 AQ 27, letter to Joubert, June 27, 1785.
16. AN, 131 MI 53 AQ 28, letter to Grand, June 19, 1786.
17. Translations of the full texts of both edicts can be found in Lalande, *Art of Papermaking*, 64–80.
18. Desmarest's analysis of the effects and evasions of the decrees of 1730 and 1741 appeared in "Papier," 547–51.
19. ADH, C 2670, "Etat des différentes sortes et qualités des papiers qui se fabriquent en Vivarais."
20. Family archive of Régis de Montgolfier, letter of August 1786.
21. For these requirements, see Lalande, *Art of Papermaking*, 65.
22. Lalande, *Art of Papermaking*, 64.
23. On early French experiments with Hollander beaters, see Charles Ballot, *L'Introduction du machinisme dans l'industrie française* (Paris: F. Rieder, 1923), 555–56; for the Montgolfiers' first, failed trials with these machines, see chs. 4–5.
24. "Mémoire du sieur Montgolfier fabricant et propriétaire d'une fabrique à Annonay."
25. Family archive of Régis de Montgolfier, letter of August 1786.
26. Lalande, *Art of Papermaking*, 69–70.
27. Quoted in Briquet, "Associations et grèves," 171.
28. Briquet, "Associations et grèves," 171.
29. For the Auvergne in general, see Pierre Claude Reynard, "La Papeterie ambertoise au XVIIIe siècle: Une Prospérité fragile et stérile" (Ph.D. diss., York University, 1994), 132; for Chamalières, see Gazel, *Anciens Ouvriers*, 40.

30. Briquet, "Associations et grèves," 176.
31. Reynard, "*Intendants, Subdélégués*," 13–14.
32. Briquet, "Associations et grèves," 176.
33. Desmarest, "Papier," 548.
34. Henry Tournier, "Syndicats ouvriers et grèves révolutionnaires dans le Castrais à la fin de l'Ancien Régime," in *Mémoires de l'Académie des sciences inscriptions et belles-lettres de Toulouse*, 12th ser., 3 (Toulouse, 1925), 237.
35. The son of a master seeking his father's rank was relieved of the responsibility of fashioning a masterpiece. On the credentials expected of aspirants to mastership, see Lalande, *Art of Papermaking*, 71–72.
36. Lalande, *Art of Papermaking*, 66, 72.
37. Cited in Boithias and Mondin, *Les Moulins à papier . . . d'Auvergne*, 55.
38. Boithias and Mondin, *Les Moulins à papier . . . d'Auvergne*, 56.
39. Iain Cameron, *Crime and Repression in the Auvergne and the Guyenne, 1720–1790* (Cambridge: Cambridge University Press, 1981), 171.
40. "Etat des moulins ou manufactures à papier situés sur la rivière de Deûme, lès Annonay."
41. Reynard, *La Papeterie ambertoise*, 87, Graph 3.
42. Reynard, *La Papeterie ambertoise*, 83, Graph 2.
43. Lalande, *Art of Papermaking*, 61.
44. Lalande, *Art of Papermaking*, 62.
45. For a brief, clear discussion of consumption during the twilight of the Old Regime, see C. Jones, "Bourgeois Revolution Revivified," 88–93.
46. On book ownership among the common people, see Daniel Roche, *The People of Paris: An Essay in Popular Culture in the Eighteenth Century*, trans. Marie Evans (Berkeley and Los Angeles: University of California Press, 1987), ch. 7, esp. Table 7.1; and on the widening uses of paper, see Sonenscher, *Work and Wages*, 200, 230.
47. Darnton, *Business of Enlightenment*, 185–96.
48. Reynard, *La Papeterie ambertoise*, 70–276.
49. "Mémoire du sieur Montgolfier fabricant et propriétaire d'une fabrique à Annonay."
50. Jan de Vries and Ad van der Woude, *The First Modern Economy: Success, Failure, and Perseverance of the Dutch Economy, 1500–1815* (Cambridge: Cambridge University Press, 1997), 313–14.
51. AN, F[12] 1479, Antoine-Laurent de Lavoisier and Jean-Baptiste Jumelin, "Rapport concernant le c[itoyen] Desmarest," May 8, 1793.
52. Quoted in Eugène Creveaux, "Un Grand Ingénieur papetier: Jean-Guillaume Ecrevisse, collaborateur de Nicolas Desmarest," in *Contribution à l'histoire de la papeterie en France*, 5 (Grenoble: Editions de "L'Industrie Papetière," 1937), 13.
53. Desmarest, "Papier," 495.
54. Lalande, *Art of Papermaking*, 28.

55. Desmarest, "Papier," 522.
56. Ecrevisse's calculation appears in Creveaux, "Ingénieur papetier," 45.
57. André, *Machines à papier*, 56.
58. Ballot, *Introduction du machinisme*, 555–56.
59. Reynaud, *Moulins à papier*, 94, including the quoted phrase.
60. Quoted in Reynaud, *Moulins à papier,* 94.
61. Antoine-François Montgolfier to the Estates of Languedoc, 1780.
62. Quoted in Deyon and Guignet, "The Royal Manufactures," 621.
63. Reynard, "*Intendants, Subdélégués*," 10–24.
64. Alexis de Tocqueville, *The Old Régime and the French Revolution*, trans. Stuart Gilbert (Garden City, N.Y.: Doubleday, 1955), 237–38, n. 17.
65. For one small producer eager to obtain Hollander beaters, see my essay "The Perils of Petty Production: Pierre and Jean-Baptiste Serve of Chamalières," *Science in Context* 11, no. 1 (1998): 3–21.
66. Pierre Bonnassieux and Eugène Lelong, *Conseil de Commerce et Bureau du Commerce, 1700–1791: Inventaire analytique des procès-verbaux* (Paris: Imprimerie nationale, 1900), xiii. On the rather complex history of the Bureau of Commerce, see Gillispie, *Science and Polity*, 388–478; Hilaire-Pérez, "Invention and the State"; Parker, *An Administrative Bureau*; and idem, *Bureau of Commerce.*
67. Formally in the charge of the Council of Commerce, the individual inspectors generally took their marching orders from the Bureau of Commerce and the provincial intendants. The most valuable recent work on the inspectors is Minard, *La Fortune du colbertisme*; idem, "L'Inspection des manufactures"; and idem, "La Fin de l'inspection des manufactures." See also André Rémond, *John Holker, manufacturier et grand fonctionnaire en France au XVIIIe siècle, 1719–1786* (Paris: Marcel Rivière, 1946).
68. Jean-Marie Roland de la Platière, "Inspecteurs," in *Encyclopédie méthodique: Manufactures, arts, et métiers*, 1 (Paris, 1785), 69.
69. Minard, "La Fin de l'inspection," 302–3.
70. On Desmarest, see my essay "Nicolas Desmarest and the Transfer of Technology in Old Regime France," in *The Modern Worlds of Business and Industry: Cultures, Technology, Labor*, ed. Karen Merrill (Turnhout: Brepols, 1998), 103–20.
71. AN, F^{12} 737, Turgot to Daniel-Charles Trudaine, September 14, 1762.
72. AN, F^{12} 1474, letter of March 14, 1788.
73. Desmarest, "Papier," 550.
74. The quoted passages are from Desmarest, "Papier," 548–49.
75. Quoted in Coleman, *British Paper Industry*, 53.
76. David Jeremy, "Damming the Flood: British Government Efforts to Check the Outflow of Technicians and Machinery, 1780–1843," *Business History Review* 51 (Spring 1977): 3, n. 2.
77. Harris, *Essays in Industry*, 87–89, 99, 169–70; Gillispie, *Science and Polity*, 428.
78. Harris, *Essays in Industry.*
79. AN, F^{12} 1479, Nicolas Desmarest, "Mémoire pour M. Desmarest."

80. Quoted in Kenneth Taylor, "Nicolas Desmarest, 1725–1815: Scientist and Industrial Technologist" (Ph.D. diss., Harvard University, 1968), 188.

81. Nicolas Desmarest, "Premier mémoire sur les principales manipulations qui sont en usage dans les papeteries de Hollande, avec l'explication physique des résultats de ces manipulations," in *Mémoires de l'Académie royale des sciences de Paris* (Paris: Imprimerie royale, 1774), 335–64; and idem, "Second mémoire sur la papeterie," in *Mémoires de l'Académie royale des sciences de Paris* (Paris: Imprimerie royale, 1778), 599–687.

82. Desmarest, "Second mémoire," 667.

83. Desmarest, "Second mémoire," 683; and Desmarest to Ecrevisse, February 29, 1776, quoted in Creveaux, "Ingénieur papetier," 21.

84. Desmarest, "Second mémoire," 667.

85. For Desmarest's extended discussion of his countrymen's "prejudices," see "Second mémoire," 670–79.

86. Desmarest's clearest comments on the relative merits and drawbacks of French and Dutch reams appear in "Papier," 525–28. Lalande's appraisals are found in *Art of Papermaking*, 57–59.

87. Desmarest, "Second mémoire," 680, 682.

88. AN, F[12] 1474, letters of July 4, July 12, and September 7, 1774.

89. Lavoisier and Jumelin, "Rapport."

90. On the deficiencies of the devices, see Desmarest, "Second mémoire," 684; on the product of these shortcomings, see "Second mémoire," 668.

91. Ecrevisse to Desmarest [1775], quoted in Creveaux, "Ingénieur papetier," 21.

92. For a full treatment of Ecrevisse's life and work, see Creveaux, "Ingénieur papetier," passim. The quoted passage appears on 24–25.

93. Desmarest, "Second mémoire," 686.

94. Desmarest, "Mémoire sur l'état actuel des papeteries d'Annonay."

95. Quoted in Reynaud, *Moulins à papier*, 85.

96. AN, 131 MI 53 AQ 25, letter of December 18, 1779.

97. Desmarest, "Mémoire sur l'état actuel des papeteries d'Annonay."

98. Desmarest, "Mémoire sur l'état actuel des papeteries d'Annonay."

99. ADA, C 143, Mathieu and Pierre-Louis Johannot to the Estates of Languedoc; Antoine-François Montgolfier to the Estates of Languedoc, 1780; Gillispie, *Science and Polity*, 453–54.

100. For the initial financial arrangements, see ADH, C 2679; on the outstanding 9,000 livres, see ADA, C 143, Etienne Montgolfier to the Estates of Languedoc, printed petition [1783 or 1784].

101. Desmarest, "Mémoire sur l'état actuel des papeteries d'Annonay."

102. AN, 131 MI 53 AQ 22, document 4.

103. Desmarest, "Mémoire pour M. Desmarest."

104. Desmarest, "Second mémoire," 686.

105. Desmarest, "Mémoire pour M. Desmarest."

106. AN, F^{12} 2281, Henri Villarmain, "Mémoire sur l'établissement d'une papeterie à cylindres en Angoumois."
107. Reynaud, *Moulins à papier*, 121.
108. Reynaud, *Moulins à papier*, 121–26.
109. Desmarest, "Papier," 548.
110. "Mémoire du sieur Montgolfier fabricant et propriétaire d'une fabrique à Annonay."
111. Desmarest, "Second mémoire," 684.
112. Léon, "La Réponse de l'industrie," 249.

CHAPTER 5: BUILDING THE BEATERS AND THE JOURNEYMEN'S CUSTOM

1. Quoted in Creveaux, "Ingénieur papetier," 14.
2. Quoted in Creveaux, "Ingénieur papetier," 45.
3. Three fine starting points on the issue of proletarianization are Christopher Johnson, "Patterns of Proletarianization: Parisian Tailors and Lodève Woolens Workers," in *Consciousness and Class Experience*, 65–84; idem, "Economic Change and Artisan Discontent: The Tailors' History, 1800–48," in *Revolution and Reaction: 1848 and the Second French Republic*, ed. Roger Price (London: Croom Helm, 1975), 87–114; and Scott, *Glassworkers of Carmaux*, 1–6, 19–52, 72–107.
4. Sonenscher, *Work and Wages*, 131–32, 140, 177–78, 185–88, 205, 209, 321–27.
5. Quoted in Briquet, "Associations et grèves," 186.
6. Sonenscher, *Work and Wages*, 174–209, esp. 176–78.
7. AN, 131 MI 53 AQ 22, document 3.
8. Etienne Montgolfier to the Estates of Languedoc, printed petition [1783 or 1784].
9. "Mémoire du sieur Montgolfier fabricant et propriétaire d'une fabrique à Annonay."
10. Quoted in Reynaud, *Moulins à papier*, 95.
11. Reynaud, *Moulins à papier*, 111.
12. Etienne Montgolfier to the Estates of Languedoc, printed petition [1783 or 1784].
13. Reynaud, *Moulins à papier*, 115.
14. ADA, C 143, letter of November 29, 1781.
15. Quoted in Reynaud, *Moulins à papier*, 53.
16. For "engeance" and similar imprecations, see Gachet, "Grèves d'ouvriers papetiers," 138.
17. L. Lescourre, quoted in Alexandre Nicolaï, *Histoire des moulins à papier du Sud-Ouest de la France*, 1 (Bordeaux: G. Delmas, 1935), 63.
18. Quoted in Gillispie, *Montgolfier Brothers*, 17.
19. Quoted in Nicolaï, *Histoire des moulins à papier*, 60.
20. Quoted in Nicolaï, *Histoire des moulins à papier*, 64.
21. Quoted in Briquet, "Associations et grèves," 178.

22. "Mémoire sur les papeteries du Dauphiné."
23. AN, F12 2368, Jean-Baptiste Serve, "Mémoire adressé à Son Excellence le Ministre des Manufactures et du Commerce concernant les abus qu'exercent les ouvriers papetiers dans les ateliers, et qu'ils nomment lois du métier" [1806?].
24. "Mémoire sur les papeteries du Dauphiné."
25. Mignot, quoted in Gachet, "Grèves d'ouvriers papetiers," 130.
26. Jubié, quoted in Briquet, "Associations et grèves," 178.
27. For these provisions of the decree of 1727, see Desmarest, "Papier," 539–40.
28. Desmarest, "Papier," 540, 545.
29. Lalande, *Art of Papermaking*, 73.
30. AN, 131 MI 53 AQ 23, document 43.
31. Boithias and Mondin, *Les Moulins à papier . . . d'Auvergne*, 223, 227–28.
32. Auguste Lacroix, *Historique de la papeterie d'Angoulême* (Paris: Ad. Lainé et J. Havard, 1863), 40.
33. AN, 131 MI 53 AQ 23, document 5.
34. AN, 131 MI 53 AQ 23, document 9.
35. "Mémoire sur les papeteries du Dauphiné."
36. On the compagnonnages, see Sonenscher, *Work and Wages*, 295–327; and Cynthia Truant, *The Rites of Labor: Brotherhoods of Compagnonnage in Old and New Regime France* (Ithaca: Cornell University Press, 1994).
37. On confraternities and laboring men in general, see Gervase Rosser, "Crafts, Guilds, and the Negotiation of Work in the Medieval Town," *Past and Present*, no. 154 (February 1997): 3–31; David Garrioch and Michael Sonenscher, "*Compagnonnages*, Confraternities, and Associations of Journeymen in Eighteenth-Century Paris," *European History Quarterly* 16 (January 1986): 25–45; and Sonenscher, *Work and Wages,* 83–86, 207–8. On the paperworkers' confraternities in the Auvergne, see Gazel, *Anciens Ouvriers*, 173–79.
38. AN, 131 MI 53 AQ 23, document 43.
39. AN, 131 MI 53 AQ 23, document 44.
40. Serve, "Mémoire."
41. AN, 131 MI 53 AQ 23, document 5.
42. Quoted in Sonenscher, *Work and Wages*, 250–51.
43. Gazel, *Anciens Ouvriers*, 188.
44. AN, 131 MI 53 AQ 23, documents 5 and 44.
45. "Mémoire sur les papeteries du Dauphiné."
46. Cameron, *Crime and Repression*, 228–29.
47. "Mémoire sur les papeteries du Dauphiné."
48. Quoted in Boithias and Mondin, *Les Moulins à papier . . . d'Auvergne*, 227.
49. "Mémoire sur les papeteries du Dauphiné."
50. On the Dauphinois journeymen, see "Mémoire sur les papeteries du Dauphiné"; on the German paperworkers, see A. Renker, "Some Curious Customs of Old-Time Papermaking in Germany," *The Paper Maker* 30:1 (1961): 6.

51. "Mémoire sur les papeteries du Dauphiné."
52. Quoted in Gachet, "Grèves d'ouvriers papetiers," 133.
53. "Mémoire sur les papeteries du Dauphiné."
54. "Mémoire sur les papeteries du Dauphiné."
55. Quoted in Gachet, "Grèves d'ouvriers papetiers," 127.
56. "Mémoire sur les papeteries du Dauphiné."
57. Quoted in Briquet, "Associations et grèves," 172.
58. André, *Machines à papier*, 51.
59. Serve, "Mémoire."
60. Gazel, *Anciens Ouvriers*, 220, n. 1.
61. For the word *"damnation,"* see Serve, "Mémoire."
62. "Mémoire sur les papeteries du Dauphiné."
63. Serve, "Mémoire."
64. "Mémoire sur les papeteries du Dauphiné."
65. Serve, "Mémoire."
66. Jubié, quoted in Gazel, *Anciens Ouvriers*, 226.
67. "Mémoire sur les papeteries du Dauphiné."
68. Briquet, "Associations et grèves," 178.
69. AN, 131 MI 53 AQ 23, document 44.
70. Briquet, "Associations et grèves," 184–85.
71. Quoted in Gachet, "Grèves d'ouvriers papetiers," 131. On the issue and depiction of insubordinate journeymen in the eighteenth-century French trades, see Kaplan, "Social Classification," and idem, "Réflexions sur la police du monde du travail, 1700–1815," *Revue historique* 261 (January–March 1979): 17–77. For a splendid case study of the journeymen printers, see Natalie Davis, "A Trade Union in Sixteenth-Century France," *Economic History Review,* 2d ser., 19 (April 1966): 48–69.
72. Quoted in Gachet, "Grèves d'ouvriers papetiers," 139.
73. Quoted in Gachet, "Grèves d'ouvriers papetiers," 136.
74. Pierre Léon, "Morcellement et émergence du monde ouvrier," in *Histoire économique et sociale de la France, 1660–1789*, 2:659, n. 1. This figure should be considered an educated guess.
75. The full text of the Council of State's decrees appears in Desmarest, "Papier," 553–55. The quoted material is from 553–54.
76. Serve, "Mémoire."
77. AN, 131 MI 53 AQ 23, document 45.
78. Quoted in Gachet, "Grèves d'ouvriers papetiers," 129.
79. Quénot, quoted in Briquet, "Associations et grèves," 189.
80. Merville, quoted in Gachet, "Grèves d'ouvriers papetiers," 129.
81. André, *Machines à papier*, 49, and n. 28 (on the same page) for the earlier claims.
82. Quoted in Gachet, "Grèves d'ouvriers papetiers," 128.
83. Gachet, "Grèves d'ouvriers papetiers," 129.

84. Gachet, "Grèves d'ouvriers papetiers," 130.
85. AN, 131 MI 53 AQ 23, document 45.
86. Mignot, quoted in Briquet, "Associations et grèves," 175.
87. "Mémoire sur les papeteries du Dauphiné."
88. Briquet "Associations et grèves," 178.
89. Quoted in Gachet, "Grèves d'ouvriers papetiers," 138.
90. Serve, "Mémoire."
91. AN, 131 MI 53 AQ 23, document 45.

CHAPTER 6: THE LOCKOUT

1. Georg Eineder, *The Ancient Paper-Mills of the Former Austro-Hungarian Empire and Their Watermarks* (Hilversum: Paper Publications Society, 1960), 133.
2. AN, F^{12} 1477, Etienne Montgolfier, letter of August 4, 1785.
3. Gachet, "Grèves d'ouvriers papetiers," 146, n. 51.
4. Quoted in Gachet, "Grèves d'ouvriers papetiers," 137.
5. Quoted in G. Schaefer and A. Latour, "The Paper Trade before the Invention of the Paper-Machine," *Ciba Review* 72 (1949): 2654.
6. Quoted in Creveaux, "Ingénieur papetier," 13–14.
7. Jacquemant, quoted in Creveaux, "Ingénieur papetier," 12.
8. Ecrevisse, quoted in Creveaux, "Ingénieur papetier," 23.
9. For Ecrevisse's role in training Vidalon's new hands, see the letter addressed to him from Annonay on May 4, 1781, in Creveaux, "Ingénieur papetier," 52–53.
10. Quoted in Reynaud, *Moulins à papier*, 53.
11. Reynaud, *Moulins à papier*, 166.
12. Reynaud, *Moulins à papier*, 166.
13. Quoted in Gazel, *Anciens Ouvriers*, 204.
14. AN, 131 MI 53 AQ 23, document 9.
15. AN, 131 MI 53 AQ 23, document 45.
16. AN, 131 MI 53 AQ 23, document 5.
17. AN, 131 MI 53 AQ 23, document 43.
18. AN, 131 MI 53 AQ 23, document 45.
19. The quoted phrase is from Gillispie, *Montgolfier Brothers*, 21.
20. Desmarest to Ecrevisse, February 29, 1776, quoted in Creveaux, "Ingénieur papetier," 21.
21. Briquet, "Associations et grèves," 178.
22. Augustin Montgolfier to the intendant of Dauphiné, March 16, 1781, quoted in Briquet, "Associations et grèves," 182. Briquet quoted only a portion of the letter.
23. AN, 131 MI 53 AQ 23, document 45.
24. Rostaing, *Famille de Montgolfier*, 118.
25. AN, 131 MI 53 AQ 23, documents 47 and 48.

26. AN, 131 MI 53 AQ 23, document 47.
27. AN, 131 MI 53 AQ 23, document 47.
28. AN, 131 MI 53 AQ 23, document 48.
29. Gillispie, *Montgolfier Brothers*, 135; and Rostaing, *Famille de Montgolfier*, 121–22.
30. Quoted in Creveaux, "Ingénieur papetier," 53, letter of May 4, 1781. After decades of attention to bread riots, the antagonisms of the atelier and manufactory during the Old Regime are receiving renewed study. Of particular note are Haim Burstin, "Problems of Work during the Terror," in *The French Revolution and the Creation of Modern Political Culture: The Terror*, 4, ed. Keith Baker (Oxford: Pergamon, 1994): 271–93; Catharina Lis and Hugo Soly, "'An Irresistible Phalanx': Journeymen Associations in Western Europe, 1300–1800," *International Review of Social History*, supplement 2 [*Before the Unions*], 39 (1994): 11–52; and Leonard Rosenband, "Jean-Baptiste Réveillon: A Man on the Make in Old Regime France," *French Historical Studies* 20 (Summer 1997): 481–510.
31. AN, 131 MI 53 AQ 23, document 45.
32. AN, 131 MI 53 AQ 23, documents 46 and 49.
33. AN, 131 MI 53 AQ 23, document 45.
34. AN, 131 MI 53 AQ 23, document 46.
35. AN, 131 MI 53 AQ 23, document 46.
36. AN, 131 MI 53 AQ 23, document 45.
37. AN, 131 MI 53 AQ 23, documents 45, 46, and 49.
38. AN, 131 MI 53 AQ 23, document 46.
39. AN, 131 MI 53 AQ 23, document 45.
40. AN, 131 MI 53 AQ 23, document 45, and Reynaud, *Moulins à papier*, 169.
41. Quoted in Reynaud, *Moulins à papier*, 170. Augustin Montgolfier also promised to supply Vidalon's *patrons* with three workers "proscribed" by the journeymen's association, but it is not clear whether they ever arrived, or even left Dauphiné (AN, 131 MI 53 AQ 23, document 47).
42. AN, 131 MI 53 AQ 23, document 45.
43. AN, 131 MI 53 AQ 27, letter of June 27, 1785.
44. AN, 131 MI 53 AQ 23, document 45.
45. Reynaud, *Moulins à papier*, 170.
46. Reynaud, *Moulins à papier*, 171.

CHAPTER 7: THE NEW REGIME

1. Quoted in David Landes, "What Do Bosses Really Do?," *Journal of Economic History* 46 (September 1986): 599.
2. Quoted in Creveaux, "Ingénieur papetier," letter of May 4, 1781, 53.
3. Darnton, *Business of Enlightenment*, 180.
4. Etienne Montgolfier to the Estates of Languedoc, printed petition [1783 or 1784].
5. AN, 131 MI 53 AQ 27, August 18, 1785, letter to Jovin.

6. AN, 131 MI 53 AQ 27, June 27, 1785, letter to Joubert.
7. AN, 131 MI 53 AQ 27, August 27, 1785, letter to Palhion.
8. AN, 131 MI 53 AQ 23, documents 1–7. Only document 3, "my ideas on the government of a paper mill where one feeds the workers," was dated (August 25, 1785). Five of the other codes detail the work of the governors of the Hollander beaters, strongly suggesting that they were written during the early 1780s or later.
9. AN, 131 MI 53 AQ 23, documents 3, 4, and 7.
10. The quoted phrase is from Darnton, *Business of Enlightenment*, 203. See also the discussion of labor turnover in Richard Goldthwaite, *The Building of Renaissance Florence: An Economic and Social History* (Baltimore: Johns Hopkins University Press, 1980), 298–301.
11. Quoted in Donald Coleman, "Combinations of Capital and of Labour in the English Paper Industry, 1789–1825," *Economica*, new ser., 21 (February 1954): 44.
12. This process is very different from that described in Braverman, *Labor and Monopoly Capital*, 3–152.
13. Pollard, *Genesis of Modern Management*, 59–60.
14. For instance, see AN, 131 MI 53 AQ 24, "Journal concernant les ouvriers et ouvrières servantes et valets commencé le 4 avril 1784" (henceforth JO), June 22, 1784.

CHAPTER 8: HIRING AND FIRING

1. AN, 131 MI 53 AQ 27, June 27, 1785, letter to Joubert.
2. AN, 131 MI 53 AQ 28, June 19, 1786, letter to Grand.
3. The quoted phrase is the title of ch. 5 of Pollard's remarkable *Genesis of Modern Management*, 160.
4. On internal labor markets, see Peter Doeringer and Michael Piore, *Internal Labor Markets and Manpower Analysis* (Lexington, Mass.: D. C. Heath, 1971).
5. AN, 131 MI 53 AQ 25, letter to Richard et fils, January 20, 1780.
6. AN, 131 MI 53 AQ 25, letter to Augustin Montgolfier, August 26, 1780.
7. For Mosnier, see AN, 131 MI 53 AQ 93, "Journal général" (henceforth JG), October 31, 1783; for Duranton, see JO, January 12, 1786.
8. AN, 131 MI 53 AQ 28, letter to Thollet, May 7, 1787.
9. Olwen Hufton, *The Poor of Eighteenth-Century France, 1750–1789* (Oxford: Clarendon Press, 1974), 15.
10. For examples, see JO, October 24, 1784, May 22, 1785, and November 17, 1787.
11. AN, 131 MI 53 AQ 28, letter to a manufacturer in Ambert, October 24, 1786.
12. JO, July 4, 1786.
13. JO, July 4, 1786.
14. JO, April 6, 1784.
15. JO, July 25, 1789.
16. JO, November 25, 1786.
17. JG, November 30, 1783 (Bourgogne); JO, April 20, 1789 (Baugi); AN, 131 MI 53 AQ 23, document 12 (Voulouzan).

18. JO, February 29, 1788.
19. JO, April 20, 1789.
20. AN, 131 MI 53 AQ 23, document 12.
21. JO, April 25, 1787, and March 22, 1788.
22. JO, April 5, 1784.
23. JO, May 29, 1785.
24. JO, March 22, 1787. When Violet and his two sons arrived at Vidalon-le-Haut with a written endorsement from Augustin Montgolfier, they were also hired on a trial footing. For the Violets' entry into the principal Montgolfier mill, see JO, June 22, 1784.
25. Merville, quoted in Briquet, "Associations et grèves," 173.
26. Quoted in Steven Kaplan, "La Lutte pour le contrôle du marché du travail à Paris au XVIIIe siècle," *Revue d'histoire moderne et contemporaine* 36 (July–September 1989): 385.
27. Gazel, *Anciens Ouvriers*, 215.
28. The ruse by Le Bon and his wife, as well as its resolution, can be traced in JO, May 7, 1787, and AN, 131 MI 53 AQ 28, letter to Thollet, May 7, 1787.
29. For the carpenter Gagneres, see AN, 131 MI 53 AQ 27, letter of May 14, 1785; and for the apprentice layman Valençon, see AN, 131 MI 53 AQ 28, letters of October 20 and November 4, 1786.
30. JO, March 7, 1788.
31. These figures are based on the accounts of hiring in JO, primarily written by Jean-Pierre Montgolfier. The last entry in this register was made on January 22, 1790. Since this calculation is restricted to journeymen, I excluded those apprentices who arrived at Vidalon-le-Haut with certificats de congé endorsed by their former employers. The youths trained at the mill immediately before the walkout of 1781 or mentioned in the roster of workers instructed in the craft by Vidalon's *patrons* after the Hollander beaters were installed (AN, 131 MI 53 AQ 23, document 14) sometimes returned with congés signed by other masters; I also excluded both these classes of workers.
32. JO, May 17, 1784.
33. Gazel, *Anciens Ouvriers*, 203, esp. n. 1.
34. JO, April 9, 1787.
35. JO, January 16, 24, and 27, 1786.
36. JO, February 2, 1788.
37. JO, October 30, 1785.
38. JO, August 20, 1785.
39. JO, October 20, 1786.
40. JO, January 16, 1786.
41. JO, October 30, 1785.
42. JO, April 24, 1785.
43. These figures were compiled from JO.
44. JO, August 5, 1786.

45. JO, October 19, 1787.
46. JO, November 14, 1784.
47. Gazel, *Anciens Ouvriers*, 121–22.
48. Quoted in Gazel, *Anciens Ouvriers*, 121.
49. JO, August 21, 1786.
50. JO, December 28, 1785.
51. AN, 131 MI 53 AQ 92, "Journal général" (henceforth JG), September 30, 1782.
52. AN, 131 MI 53 AQ 28, letter to Jubier, January 2, 1787.
53. JO, November 6, 1785.
54. JO, September 17, 1784, and March 22, 1788.
55. JO, November 25, 1785.
56. JO, March 19, 1786.
57. JO, October 19 and 23, 1786.
58. JO, June 22, 1784.
59. JO, April 14, 1786.
60. JO, June 22, 1784.
61. JO, October 30, 1785.
62. JO, January 16, 1786, and March 22, 1788.
63. JG, June 30, 1783 (Rivoire); JO, December 2, 1785 (Charbellet).
64. JO, September 29, 1784.
65. JO, October 17, 1789.
66. JO, October 4, 1787.
67. JO, October 1, 1786.
68. JO, March 7, 1788.
69. JO, January 24, 1785.
70. JO, February 8, 1786.
71. JO, October 22, 1787.
72. JO, May 24, 1784.
73. JO, July 2, 1786.
74. JO, November 30, 1785.
75. AN, 131 MI 53 AQ 27, letter to Thollet, May 14, 1785.
76. AN, 131 MI 53 AQ 28, letter to Thollet, October 20, 1786.
77. AN, 131 MI 53 AQ 28, letter to Thollet, November 4, 1786.
78. JO, January 28, 1787.
79. JO, May 11, 1785.
80. JO, August 21, 1785.
81. JO, October 4, 1787.
82. JG, April 30 and May 31, 1781.
83. JO, August 16, 1786.
84. JO, February 8, 1786.

85. JO, October 30, 1786.
86. JO, February 4 and March 17, 1787.
87. JO, July 2, 1785.
88. JO, May 17 and 20, 1785.
89. Kaplan, "Luxury Guilds," 292, and idem, "Réflexions sur la police," 56–57.
90. Gazel, *Anciens Ouvriers*, 120.
91. JO, August 7, 1786.

CHAPTER 9: PATERNALISM

1. Quoted in Reynaud, *Moulins à papier*, 53. On industrial paternalism, see Sanford Jacoby, ed., *Masters to Managers: Historical and Comparative Perspectives on American Employers* (New York: Columbia University Press, 1991); Donald Reid, "Industrial Paternalism: Discourse and Practice in Nineteenth-Century French Mining and Metallurgy," *Comparative Studies in Society and History* 27 (October 1985): 579–607; Gérard Noiriel, "Du 'patronage' au 'paternalisme': La restructuration des formes de domination de la main-d'oeuvre ouvrière dans l'industrie métallurgique française," *Le Mouvement social* 144 (July–September 1988): 17–35; and Peter Stearns, *Paths to Authority: The Middle Class and the Industrial Labor Force in France, 1820–48* (Urbana: University of Illinois Press, 1978).
2. Quoted in Nicolaï, *Histoire des moulins à papier*, 60. For a similar statement from Vidalon-le-Haut, see AN, 131 MI 53 AQ 23, document 45: "the misery of the workers who consume all their wages *en débauche*."
3. AN, 131 MI 53 AQ 23, document 5.
4. AN, 131 MI 53 AQ 23, document 6.
5. A good starting point on the mill village is Pollard, *Genesis of Modern Management*, 197–208. See also Hamon and Perrin, *Au Coeur du XVIIIe siècle industriel*.
6. AN, 131 MI 53 AQ 23, document 6.
7. JO, March 5, 1785.
8. Hufton, *Poor of Eighteenth-Century France*, 48.
9. Marie-Hélène Reynaud, "The Daily Life in the Paper Mills of the Montgolfier just before the Revolution," in *Papers of the Twentieth International Congress of Paper Historians* (Darmstadt: Eduard Roether KG, 1990): 130.
10. Lacroix, *Historique de la papeterie*, 37–38.
11. JO, April 25, 1787.
12. JO, October 24, 1785.
13. JO, November 19, 1787.
14. JO, October 7, 1786, and March 8, 1788.
15. JG, October 30, 1779.
16. JO, September 10, 1784, September 10, 1787, and March 23, 1788.
17. JO, February 24, 1788.
18. JG, April 30, 1783.

19. JO, December 28, 1784.
20. Once the Montgolfiers passed an apprenticeship contract, however, they subtracted 60 livres from the earnings of those youngsters who failed to complete their terms. Still, even this potential shackle looks considerably less solid in light of the incentive payments made to François Millot and Jean Texier during their apprenticeships (JO, October 4, 1787). For both the penalty and the incentives, see the next chapter.
21. JO, May 11, 1785.
22. AN, 131 MI 53 AQ 22, document 3.
23. AN, 131 MI 53 AQ 22, document 3, and "Mémoire du sieur Montgolfier fabricant et propriétaire d'une fabrique à Annonay."
24. Reynaud, *Moulins à papier*, 45.
25. Reynaud, "Daily Life," 132.
26. Boithias and Mondin, *Les Moulins à papier . . . d'Auvergne*, 211.
27. JG, December 31, 1781.
28. Quoted in Alexandre Nicolaï, "La Papeterie en Angoumois au XVIIIe siècle," in *Contribution à l'histoire de la papeterie en France*, 10 (Grenoble: Editions de "L'Industrie Papetière," 1945), 15.
29. Nicolaï, "La Papeterie en Angoumois au XVIIIe siècle," 16–17, quote on 17.
30. Nicolaï, "La Papeterie en Angoumois au XVIIIe siècle," 20.
31. Quoted in Gachet, "Grèves d'ouvriers papetiers," 132.
32. Quoted in Rostaing, *Famille de Montgolfier*, 123.
33. Reynaud, "Daily Life," 132.
34. Rostaing, *Famille de Montgolfier*, 123. The Montgolfiers spoke of a *"pinte"* of wine, which, according to former usage, amounted to nine tenths of a liter.
35. AN, 131 MI 53 AQ 23, document 41.
36. Quoted in Rostaing, *Famille de Montgolfier*, 123.
37. JO, April 6, 1784.
38. AN, 131 MI 53 AQ 23, document 31.
39. Quoted in Gachet, "Grèves d'ouvriers papetiers," 138.
40. JG, March 1, 1785.
41. JG, February 28, 1786.
42. AN, 131 MI 53 AQ 23, document 5.
43. AN, 131 MI 53 AQ 23, documents 6 and 7. Evidently, the Montgolfiers had permitted their workers to nourish themselves just one month before the walkout of 1781. On September 25, the workers asked to feed themselves and, three days later, were doing so (JG, September 30, 1781). Perhaps Vidalon's *patrons* mothballed this policy once the boisterous veterans moved on and restored the master's table as another element in the training of their new men. Although the journeymen apparently had pressed for the disappearance of this customary fixture of their trade, one still wonders if this abrupt change contributed to the veterans' tenacious effort to get the bons élèves to fork over apprenticeship fees.

44. JO, December 18, 19, and 23, 1785.
45. JO, December 18, 1785; and AN, 131 MI 53 AQ 23, document 8.
46. AN, 131 MI 53 AQ 23, document 7.
47. AN, 131 MI 53 AQ 23, document 6.
48. JO, May 29, 1788.
49. JO, December 2, 1785.
50. JO, November 22, 1784.
51. JO, June 10, 1787.
52. JO, November 26, 1784.
53. AN, 131 MI 53 AQ 27, April 8, 1784.
54. JO, November 3, 1789.
55. JO, November 11, 1787.
56. JO, November 30, 1784.
57. JO, August 8, 1787.
58. AN, 131 MI 53 AQ 23, document 6; and JO, May 7, 1787.
59. AN, 131 MI 53 AQ 23, documents 6 and 47.
60. JO, December 24, 1785, and April 25, 1787.
61. JO, August 30, 1788, and February 20, 1789.
62. JO, August 21, 1786.
63. JO, May 16 and August 21, 1784.
64. JO, April 6, 1784.
65. JO, May 21, 1787.
66. JO, August 8, 1787.
67. JO, April 30, 1785, September 10, 1787, and November 5, 1789.
68. JO, December 14, 1785.
69. JO, June 17, 1787.
70. JO, November 26, 1784.
71. JO, December 18, 1786.
72. JO, February 4, 1785.
73. JO, November 24, 1784.
74. JO, August 5, 1784.
75. JO, December 20, 1786.
76. JO, December 24, 1786.
77. JO, November 27, 1785.
78. JO, August 22 and September 4, 1784.
79. JO, December 27, 1784.
80. JO, January 6, 1785.
81. JO, September 30, 1784.
82. JO, October 15, 1784, and April 7, 1789.
83. JO, May 30, 1786.

84. JO, June 7, 1786.
85. JO, February 20, 1785.
86. JO, April 6, 1784.
87. AN, 131 MI 53 AQ 23, document 6.
88. JG, October 31, 1783.
89. JO, December 2, 1785.
90. JO, April 21, 1789.
91. JO, May 1, 2, and 3, 1787.
92. AN, 131 MI 53 AQ 23, document 48.
93. AN, 131 MI 53 AQ 23, document 6.
94. AN, 131 MI 53 AQ 23, documents 4, 5, and 6.
95. AN, 131 MI 53 AQ 23, document 4.
96. AN, 131 MI 53 AQ 23, document 6.

CHAPTER 10: WAGES

1. AN, 131 MI 53 AQ 23, document 47.
2. Quoted in Peter Mathias, *The Transformation of England: Essays in the Economic and Social History of England in the Eighteenth Century* (New York: Columbia University Press, 1979), 148.
3. Quoted in Michael Sonenscher, "Work and Wages in Paris in the Eighteenth Century," in *Manufacture in Town and Country before the Factory*, 150.
4. Quoted in Mathias, *Transformation of England*, 163. See also A. W. Coats, "Changing Attitudes to Labour in the Mid-Eighteenth Century," *Economic History Review* 2d ser., 11, no. 1 (1958): 35–51.
5. Quoted in Harry Payne, *The Philosophes and the People* (New Haven: Yale University Press, 1976), 118.
6. Mathias, *Transformation of England*, 161.
7. Quoted in Creveaux, "Ingénieur papetier," 10.
8. Even before the lockout, Vidalon's *patrons* offered incentives to bound men. In June 1779, they granted a premium to an "apprentice layman" who had toiled with the unwieldy paper known as *grand aigle* (JG, June 30, 1779).
9. Briquet, "Associations et grèves," 167.
10. AN, 131 MI 53 AQ 23, document 1.
11. Briquet, "Associations et grèves," 163; Lalande, *Art of Papermaking*, 71.
12. AN, 131 MI 53 AQ 23, document 48.
13. Gachet, "Grèves d'ouvriers papetiers," 143, n. 9.
14. JO, April 6, 1784. Augustin Montgolfier, who held his apprentices to a three-year term, also halved the bonus he paid for the completion of these indentures—150 livres (AN, 131 MI 53 AQ 23, document 48).
15. JO, September 11, 1784.
16. JO, August 25, 1785.

17. JO, September 4, 1784, and May 3, 1787.
18. JO, November 30, 1787.
19. For one notable example, see JO, November 18, 1786.
20. JG, March 31, 1784. Whereas the terms of female apprenticeship in Auvergnat papermaking were evidently subject to the needs of individual workers and their masters, the Montgolfiers typically followed a more systematic approach. For the Auvergne, see Gazel, *Anciens Ouvriers*, 67–68. In 1785, Vidalon's *patrons* engaged Marguerite Brunel as an "apprentice paperworker" (JO, April 17, 1785). When Colombe Morelle moved on, she had completed her apprenticeship and been in the Montgolfiers' service for three and a half years (JO, July 31, 1785). Marie Valençon, "having made her apprenticeship" at the mill, left after four years there (JO, October 1, 1786). It is reasonable to assume that female apprenticeship at Vidalon-le-Haut lasted half as long as the term required of men—doubtless a measure of the relative value the Montgolfiers placed on the skills of the ouvriers and the time necessary to master them.
21. JO, May 3, 1787.
22. JO, June 22, 1788.
23. JO, December 26, 1786.
24. JO, July 25, 1789.
25. JO, March 9, 1788.
26. Lalande, *Art of Papermaking*, 71.
27. JO, November 18, 1786.
28. JO, February 9, 1786.
29. JO, June 3, 1786.
30. JO, April 6, 1784, and July 14, 1785.
31. AN, 131 MI 53 AQ 23, document 10.
32. JO, February 9, 1786.
33. JO, November 10, 1786.
34. JO, February 14, 1785.
35. JO, January 15, 1789.
36. JO, August 18, 1788.
37. JO, October 5, 1787.
38. JG, July 31, 1783; JO, July 30, 1785, and March 17, 1787.
39. For England, see Coleman, *British Paper Industry*, 298–300; for France, see Desmarest, "Papier," 511, Table.
40. Eric Hobsbawm, *Labouring Men: Studies in the History of Labour* (London: Weidenfeld and Nicolson, 1964), 353.
41. AN, 131 MI 53 AQ 23, document 16.
42. Archives des papeteries Canson et Montgolfier, Annonay, uncatalogued. A copy of this document is in the author's possession.
43. AN, 131 MI 53 AQ 23, document 32; and JO, October 17, 1789.

44. Desmarest, "Papier," 556.
45. Lalande, *Art of Papermaking*, 73.
46. JG, July 31, 1780.
47. JO, April 18, 1786.
48. JG, July 31, 1780.
49. JO, March 17, 1787.
50. AN, 131 MI 53 AQ 23, document 8.
51. AN, 131 MI 53 AQ 23, document 10.
52. JG, July 31, 1780.
53. Briquet, "Associations et grèves," 187.
54. JO, October 4, 1787.
55. Rosenband, "Réveillon," 500.
56. Léon, "Morcellement et émergence du monde ouvrier," 669–70.
57. JO, December 22, 1787.
58. Gazel, *Anciens Ouvriers*, 125–27.
59. AN, 131 MI 53 AQ 23, document 32.
60. AN, 131 MI 53 AQ 23, document 1.
61. JO, October 5, 1785.
62. JO, February 9, 1786.
63. JG, December 31, 1781.
64. JG, March 31, 1781.
65. JG, December 31, 1781.
66. JG, February 10, 1782.
67. JG, February 10, 1782, and July 10, 1782.
68. AN, 131 MI 53 AQ 23, document 48.
69. Pollard, *Genesis of Modern Management*, 161.
70. JO, April 27, 1786.
71. JO, November 6, 1786.

CHAPTER 11: DISCIPLINE

1. Richard Herring, *Paper and Paper Making, Ancient and Modern*, 3d ed. (London: Longman, 1863), 53.
2. Sewell, "Visions of Labor," 258–86. Also note the comments of Darnton, *Business of Enlightenment*, 242, n. 98.
3. Whereas Taylor and his associates intended to reconfigure work (and skill) and thereby also alter workshop culture, the Montgolfiers wished to oust the journeymen's culture in order to take command of the production and transmission of skill—while leaving papermaking's fundamental division of labor and skills intact. An elegant summary of Taylorism and its American application can be found in David Montgomery, *The Fall of the House of Labor: The Workplace, the State, and*

American Labor Activism, 1865–1925 (Cambridge: Cambridge University Press, 1987), ch. 5.

4. Franco Venturi, *Italy and the Enlightenment: Studies in a Cosmopolitan Century*, trans. Susan Corsi (New York: New York University Press, 1972), xix–xx.
5. AN, 131 MI 53 AQ 28, letter of October 24, 1786.
6. AN, 131 MI 53 AQ 23, document 3.
7. Stanley Chapman and Serge Chassagne, *European Textile Printers in the Eighteenth Century: A Study of Peel and Oberkampf* (London: Heinemann Educational Books, 1981), 177.
8. Pollard, *Genesis of Modern Management*, 183.
9. AN, 131 MI 53 AQ 23, documents 5 and 17.
10. AN, 131 MI 53 AQ 23, document 4.
11. AN, 131 MI 53 AQ 23, document 6.
12. JO, January 30, February 8, and February 15, 1786.
13. JO, May 24, May 26, and September 29, 1784.
14. AN, 131 MI 53 AQ 23, document 4; the same concern was voiced in document 5.
15. Briquet, "Associations et grèves," 167.
16. Boithias and Mondin, *Les Moulins à papier . . . d'Auvergne*, 212.
17. AN, 131 MI 53 AQ 23, document 5; and Reynaud, "Daily Life," 129.
18. AN, 131 MI 53 AQ 23, document 16.
19. AN, 131 MI 53 AQ 23, document 10.
20. AN, 131 MI 53 AQ 23, document 1.
21. AN, 131 MI 53 AQ 23, document 5; and JO, April 14, 1786.
22. Gazel, *Anciens Ouvriers*, 71.
23. ADA, Q 527.
24. André, *Machines à papier*, 46; Boithias and Mondin, *Les Moulins à papier . . . d'Auvergne*, 224; Briquet, "Associations et grèves," 177; and Gazel, *Anciens Ouvriers*, 71.
25. Lalande, *Art of Papermaking*, 73.
26. Briquet, "Associations et grèves," 177; and Gachet, "Grèves d'ouvriers papetiers," 130.
27. AN, 131 MI 53 AQ 23, document 5.
28. AN, 131 MI 53 AQ 23, document 39.
29. AN, 131 MI 53 AQ 23, document 19, "pour la lumière."
30. JG, October 31, 1783; and JO, September 22, 1786.
31. AN, 131 MI 53 AQ 23, document 17.
32. AN, 131 MI 53 AQ 23, document 7.
33. JO, December 2, 1785.
34. AN, 131 MI 53 AQ 91, "Journal général" (henceforth JG), February 1, 1778.
35. JO, September 7, 1785.
36. JO, August 25, 1787.

37. AN, 131 MI 53 AQ 23, document 39.
38. JO, November 3, 1789.
39. AN, 131 MI 53 AQ 23, document 39.
40. Thompson, *Customs in Common*, 354.
41. AN, 131 MI 53 AQ 23, document 40.
42. JO, January 19, 1787 (work), and March 15, 1789 (watch).
43. AN, 131 MI 53 AQ 23, document 7.
44. AN, 131 MI 53 AQ 23, document 24.
45. AN, 131 MI 53 AQ 23, document 7.
46. AN, 131 MI 53 AQ 23, documents 4, 5, 6, and 7.
47. AN, 131 MI 53 AQ 23, document 7.
48. JO, June 1, 1789.
49. AN, 131 MI 53 AQ 23, document 7.
50. AN, 131 MI 53 AQ 23, document 4.
51. AN, 131 MI 53 AQ 23, document 6.
52. F. Meissner, "Some Notes about a Papermakers' Association in Slovakia during the Eighteenth Century," *The Paper Maker* 39:2 (1960): 44.
53. AN, 131 MI 53 AQ 23, document 7.
54. AN, 131 MI 53 AQ 23, document 3.
55. AN, 131 MI 53 AQ 23, document 17.
56. AN, 131 MI 53 AQ 23, document 4.
57. AN, 131 MI 53 AQ 23, document 6.
58. JO, July 20, 1787.
59. AN, 131 MI 53 AQ 23, document 3.
60. JO, September 22, 1784.
61. JO, August 10, 1784.
62. JO, October 17, 1785.
63. JO, October 10, 1785.
64. JO, October 7, 1784.
65. JO, February 23, 1788.
66. AN, 131 MI 53 AQ 266, June 1804 (Bonnetton); and JO, January 28, 1787.
67. AN, 131 MI 53 AQ 265, February 1800.
68. JO, May 16, 1784.
69. AN, 131 MI 53 AQ 23, document 17.
70. AN, 131 MI 53 AQ 23, document 17.
71. JO, May 24, 1785.
72. AN, 131 MI 53 AQ 23, document 7. As a means of avoiding the design and implementation of extensive disciplinary regimes, subcontracting was widely practiced by early French and English industrialists. On this point, see Perrot, "Three Ages of Industrial Discipline," 155, and Pollard, *Genesis of Modern Management*, 189.

73. AN, 131 MI 53 AQ 23, document 1.
74. AN, 131 MI 53 AQ 23, document 3.
75. AN, 131 MI 53 AQ 23, document 3.
76. AN, 131 MI 53 AQ 23, document 7.
77. AN, 131 MI 53 AQ 23, document 4.
78. AN, 131 MI 53 AQ 23, document 3.
79. JO, January 24, 1785.
80. JO, September 28 and October 1, 1785.
81. JO, July 27, 1784.
82. JO, January 24, 1785.
83. JO, May 11, 1785.
84. JO, December 2, 1785.
85. JO, October 4, 1787.
86. Reynaud, *Moulins à papier*, 109.
87. Quoted in Creveaux, "Ingénieur papetier," letter of April 13, 1781, 52.
88. Quoted in Creveaux, "Ingénieur papetier," letter of May 4, 1781, 53.
89. AN, 131 MI 53 AQ 27, letter to Joubert, June 27, 1785.
90. Quoted in Pollard, *Genesis of Modern Management*, 174.
91. Michel Foucault, *Discipline and Punish: The Birth of the Prison*, trans. Alan Sheridan (New York: Pantheon, 1977).
92. JO, April 30, 1784, October 30, 1786, and April 8, 1787.
93. JO, September 7, 1785.
94. Steven Kaplan, *The Bakers of Paris and the Bread Question, 1700–1775* (Durham: Duke University Press, 1996), 56.
95. For Gutman's approach, see Ira Berlin's appreciation, "Introduction: Herbert G. Gutman and the American Working Class," in Herbert Gutman, *Power and Culture: Essays on the American Working Class*, ed. Ira Berlin (New York: Pantheon, 1987), 69.
96. JO, September 3, 1785.
97. AN, 131 MI 53 AQ 23, document 19 *bis*.
98. JO, November 13, 1784.
99. JO, August 14, 1786.
100. JO, February 20, 1789.
101. Sonenscher, *Work and Wages*, 131–33, 140, 177–78, 187–88.

CHAPTER 12: TECHNOLOGICAL TRANSFER

1. Pollard, *Genesis of Modern Management*, 167.
2. See the letter from Annonay of May 4, 1781, in Creveaux, "Ingénieur papetier," 53; and Reynaud, *Moulins à papier*, 110.
3. Musée de l'Air (VIII, 32), abbé Alexandre-Charles Montgolfier to Etienne Montgolfier, September 4, 1783.

4. Creveaux, "Ingénieur papetier," 67.
5. Desmarest's estimate of the scale of the French industry appears in Gachet, "Grèves d'ouvriers papetiers," 136.
6. Léon, "La Réponse de l'industrie," 249.
7. Desmarest, "Mémoire pour M. Desmarest."
8. André, *Machines à papier*, 58.
9. André, *Machines à papier*, 58.
10. Gachet, "Grèves d'ouvriers papetiers," 136.
11. AN, 131 MI 53 AQ 25, letter to Ravizé in Montpellier, September 18, 1780.
12. AN, 131 MI 53 AQ 25, letter to Falquier in Bordeaux, August 12, 1780.
13. AN, 131 MI 53 AQ 25, letter to Tavernier in Lyons, May 8, 1780.
14. AN, 131 MI 53 AQ 25, letter to Jacques Montgolfier in Paris, August 3, 1780.
15. AN, 131 MI 53 AQ 25, letter to Aublet in Paris, July 10, 1780.
16. AN, 131 MI 53 AQ 25, letter to Philis in Montpellier, August 10, 1780.
17. Quoted in Reynaud, *Moulins à papier*, 110.
18. Etienne Montgolfier to Jean-Pierre or Alexandre-Charles Montgolfier, October 23, 1783. The original document is housed in a private Montgolfier archive. A copy can be found in the Firestone Library of Princeton University.
19. Musée de l'Air (XII, 23), Jean-Baptiste Réveillon to Etienne Montgolfier, June 8, 1785.
20. David, "Mon Voyage de 1787."
21. Etienne Montgolfier to the Estates of Languedoc, printed petition [1783 or 1784].
22. David, "Mon Voyage de 1787."
23. Musée de l'Air (VIII, 32), Alexandre-Charles Montgolfier to Etienne Montgolfier, September 4, 1783.
24. The quoted phrase, from Breton papermaking, appears in Gachet, "Grèves d'ouvriers papetiers," 137.
25. David, "Mon Voyage de 1787."
26. David, "Mon Voyage de 1787."
27. On the new building at Faya, see Gillispie, *Montgolfier Brothers*, 128; for Mathieu Johannot's comment, see Reynaud, *Moulins à papier*, 129.
28. Reynaud, *Moulins à papier*, 234, 236–39 (including graphs).
29. Etienne Montgolfier to Jean-Pierre Montgolfier, November 15, 1783. The original document is housed in a private Montgolfier archive. A copy can be found in the Firestone Library of Princeton University.

CHAPTER 13: PERSISTENCE

1. This figure was compiled from the accounts in JO.
2. JO, November 30, 1785, and November 18, 1786.
3. AN, 131 MI 53 AQ 23, document 48.
4. Quoted in Creveaux, "Ingénieur papetier," 13.

5. These figures were compiled from AN, 131 MI 53 AQ 23, document 14.
6. AN, 131 MI 53 AQ 23, document 48.
7. Ecrevisse's account of his tactics at Cuvelier's mill in Flanders appears in Creveaux, "Ingénieur papetier," 13–14.
8. AN, F^{12} 1484.
9. The "Journal concernant les ouvriers et ouvrières servantes et valets commencé le 4 avril 1784" provided the data for the years 1784 through 1789, as well as much valuable information on the Vidalon careers of the new men before April 1784. AN, 131 MI 53 AQ 23, documents 13 and 14, supplied evidence about the appearance of many of the new men in the mill. AN, 131 MI 53 AQ 92, "Journal général," provided useful material for 1782–83. Etienne Montgolfier's assessment of Vidalon's entire work force in 1786 appeared in his letter from Annonay of August 1786 (family archive of Régis de Montgolfier).
10. Chassagne, *Oberkampf*, 245, Table 15.
11. Pollard, *Genesis of Modern Management*, 176.
12. Daniel Roche, "Commentary: Jacques-Louis Ménétra: An Eighteenth-Century Way of Life," afterword of Jacques-Louis Ménétra, *Journal of My Life*, ed. Daniel Roche, trans. Arthur Goldhammer (New York: Columbia University Press, 1986), 285.
13. Quoted in Léon, "Morcellement et émergence du monde ouvrier," 660.
14. AN, 131 MI 53 AQ 23, document 14, and JO, April 14 and November 6, 1786.
15. These migratory patterns were reconstructed from the written discharges carried by the tramping men hired by Vidalon's *patrons* from April 1784 through December 1789 (JO). Although they bear witness to a regional rather than a national labor market, they do suggest why the state feared a broad combination among the paperworkers. Equally, the journeymen's wide movements indicate why the Montgolfiers demanded effective state intervention in the labor market and complained that no individual manufacturer had either the time or money to police it on his own (AN, 131 MI 53 AQ 23, document 45).
16. JO, October 13, 1784.
17. JO, July 30, 1785, and March 6, 1788.
18. JO, April 25, 1787.
19. This figure was compiled from the dates and locations inscribed on the journeymen's discharges (according to the Montgolfiers) and the dates of their arrival and hiring at Vidalon-le-Haut (JO).
20. AN, F^{12} 1483.
21. JO, August 21, 1784.
22. JO, January 12, 1785.
23. JO, May 11, 1785.
24. AN, F^{12} 1483.
25. For Thiers, see AN, F^{12} 1484; for Vidalon-le-Haut and the Johannot mill (Faya-lès-Annonay), see ADA, L 803, documents 25 and 26, respectively.
26. These figures were compiled from the accounts in JO. In order to maintain statis-

tical consistency, I excluded from this group the new men who returned to Vidalon-le-Haut with references endorsed by other manufacturers.

27. This figure was compiled from ADA, L 803, document 26.
28. JO, April 24, 1785.
29. "Etat des moulins à papier."
30. JO, December 4, 1785, and February 21, 1786.
31. "Etat des moulins à papier."
32. For winter problems at Tence, see "Etat des moulins à papier." The four workers were Chêne (one day), Mercier (three months), Antoine Duranton (seven months), and Blaise Duranton (eight months). These figures were compiled from JO.
33. "Mémoire sur les papeteries du Dauphiné."
34. The quoted phrase is from Goldthwaite, *Building of Renaissance Florence*, 301; see also Christopher Hill, "Pottage for Freeborn Englishmen: Attitudes to Wage Labour in the Sixteenth and Seventeenth Centuries," in *Socialism, Capitalism, and Economic Growth*, ed. C. H. Feinstein (Cambridge: Cambridge University Press, 1967), 338–50.
35. Robert Darnton, *The Great Cat Massacre and Other Episodes in French Cultural History* (New York: Basic Books, 1984), 81.

CHAPTER 14: ATTITUDES

1. Darnton, *Great Cat Massacre*, 81.
2. Sonenscher, *Work and Wages*, 177, 187–88.
3. "Mémoire sur les papeteries du Dauphiné."
4. AN, 131 MI 53 AQ 23, documents 47 and 48.
5. "Mémoire sur les papeteries du Dauphiné."
6. These figures were compiled from JO. All of the numbers in this section refer to male hands only.
7. JO, May 29, 1785.
8. JO, February 21 and November 6, 1786, October 29, 1787, and June 22, 1788.
9. JO, January 16 and October 26, 1786.
10. JO, April 8, 1787.
11. JO, October 10, 1786, and February 20, 1789.
12. JO, May 20, 1786.
13. JO, March 21 and May 9, 1786, May 21, 1787, and October 17, 1789.
14. These figures were compiled from JO.
15. AN, 131 MI 53 AQ 23, document 3.
16. JO, October 7, 1784, October 26, 1786, October 22, 1787, and February 20, 1789.
17. JO, October 10, 1786.
18. JO, May 7, 1787.
19. JO, November 6, 1786, and February 2, 1788.

20. JO, August 8, 1787.
21. JO, April 14, 1786.
22. JO, April 2, 1786.
23. JO, June 23 and August 8, 1787.
24. AN, 131 MI 53 AQ 23, document 44.
25. JO, July 10, 1787, and February 2, 1788.
26. JO, June 9, 1786.
27. JO, February 21, 1786.
28. JO, May 21, 1787.
29. AN, 131 MI 53 AQ 23, document 47.
30. AN, 131 MI 53 AQ 23, document 6.
31. These figures were compiled from JO. For Grangeon, see JO, June 26, 1784.
32. JO, March 1, 1785.
33. JO, April 4, 1784.
34. JO, March 1, 1785.
35. JO, March 2, 1785.
36. JO, April 11, 1785.
37. JO, August 21, 1785.
38. JO, October 3, 1786.
39. JO, August 8, 1787.
40. JO, February 21, 1786.
41. JO, November 6, 1786.
42. JO, January 24, 1785.

CHAPTER 15: PRODUCTIVITY

1. Pollard, *Genesis of Modern Management*, 161–62.
2. On the history of early systems of labor discipline, see Clive Behagg, "Controlling the Product: Work, Time, and the Early Industrial Workforce in Britain, 1800–1850," in *Worktime and Industrialization: An International History*, ed. Gary Cross (Philadelphia: Temple University Press, 1988), 41–58; Cottereau, "Distinctiveness of Working-Class Cultures in France," 111–35; Robert Davis, *Shipbuilders of the Venetian Arsenal: Workers and Workplace in the Preindustrial City* (Baltimore: Johns Hopkins University Press, 1991); Peter Linebaugh, *The London Hanged: Crime and Civil Society in the Eighteenth Century* (Cambridge: Cambridge University Press, 1992), 396–401; Jan Materné, "Social Emancipation in European Printing Workshops before the Industrial Revolution," in *The Workplace before the Factory: Artisans and Proletarians, 1500–1800*, ed. Thomas Safley and Leonard Rosenband (Ithaca: Cornell University Press, 1993), 204–24; Neil McKendrick, "Josiah Wedgwood and Factory Discipline"; Perrot, "Three Ages of Industrial Discipline"; Pollard, *Genesis of Modern Management*, 160–208; John Rule, *The Experience of Labour in Eighteenth-Century English Industry* (New York: St. Martin's Press, 1981); and Thompson, *Customs in Common*, 370–403.

For two richly textured sources, see "Naissance de la classe ouvrière en France" (special issue of *Le Mouvement social*); and M. W. Flinn, ed., *The Law Book of the Crowley Ironworks* (Surtees Society, 167, 1957).

3. Douglas Reid, "The Decline of Saint Monday, 1766–1876," *Past and Present,* no. 71 (May 1976): 76–101.
4. For France, see Desmarest, "Papier," 510; for Britain, see *The Statutes at Large*, ed. Danby Pickering, 40 (Cambridge), 814. It is worth reiterating that work at most mills actually lasted only as long as the pulp did.
5. André, *Machines à papier*, 47.
6. AN, 131 MI 53 AQ 265 and 266.
7. Coleman, *British Paper Industry*, 298.
8. Jean-Marie Janot, *Les Moulins à papier de la région vosgienne*, 1 (Nancy: Berger-Levrault, 1952), 83. These arrangements appear in a contract dated 1652. The breach day was also respected at the Johannots' mill at Faya-lès-Annonay (AN, 131 MI 53 AQ 23, document 31).
9. Quoted in Léon, "Morcellement et émergence du monde ouvrier," 660.
10. Quoted in Coleman, *British Paper Industry*, 163, n. 2.
11. Lalande, *Art of Papermaking*, 60.
12. André, *Machines à papier*, 46.
13. George Rudé, *The Crowd in the French Revolution* (Oxford: Oxford University Press, 1959), 251, Appendix VII, Table 1, n. 2.
14. The English holiday calendar is from R. Johnson, "The Paper-Maker and Stationers' Assistant" (London, 1794), which is reproduced in *Dictionary and Encyclopaedia of Paper and Paper-making*, 127.
15. André, *Machines à papier*, 46–48. Partly on the basis of an early essay by the present author, André emphasizes the gap between the manufacturers' complaints about irregular output and lengthy holiday disruptions, and actual production.
16. Lalande, *Art of Papermaking*, 57.
17. "Etat des moulins à papier."
18. Vissier's output was compiled from JG, 1761.
19. AN, 131 MI 53 AQ 23, document 6.
20. On the basis of Venetian arsenal workers, Cornish tin miners, Whickham pitmen, and Vidalon-le-Haut, Robert Duplessis has observed that "by the eighteenth century, at least, there is mounting evidence that the duration, regularity, and intensity of labor were on the rise." See Robert Duplessis, *Transitions to Capitalism in Early Modern Europe* (Cambridge: Cambridge University Press, 1997), 259–309, esp. 264, 296–97.
21. Carlo Cipolla, *Before the Industrial Revolution: European Society and Economy, 1000–1700*, 2d ed. (New York: Norton, 1980), 129.

CHAPTER 16: THE HIERARCHY OF VATS

1. Pollard, *Genesis of Modern Management*, 186–92.
2. AN, 131 MI 53 AQ 23, document 47.

3. JG, July 1, 1777.
4. JG, 1761.
5. AN, 131 MI 53 AQ 265.
6. JO, October 17, 1789.
7. In conformity with the Montgolfiers' practice, I have divided movement up and down the ladder of skilled tasks into two categories, permanent and temporary transfers. This discussion is restricted to the permanent promotions and demotions. These shifts altered the team rosters that Vidalon's *patrons* placed at the head of the columns of output in their livre de fabrication (AN, 131 MI 53 AQ 265 and 266). Temporary transfers, marked within the columns of output, did not change the rosters at the top of each column.
8. JG, letter to the Prévost sisters in Paris, September 10, 1778.
9. JO, April 4, 1784.
10. JO, January 24, 1786.
11. JG, letter to Augustin Montgolfier in Rives, April 24, 1780.
12. JO, November 14, 1784.
13. Serve, "Mémoire."
14. Quoted in Gazel, *Anciens Ouvriers*, 227.

CHAPTER 17: THE FRENCH REVOLUTION AND THE PAPERMAKING MACHINE

1. On the papermakers' flush times during the first years of the French Revolution, see Darnton, *Business of Enlightenment*, 490.
2. ADA, L 803, document 2. On the Revolution in the Montgolfiers' vicinity, see abbé Charles Jolivet, *La Révolution dans l'Ardèche, 1788–1795* (Largentière: E. Mazel, 1930).
3. ADA, L 803, "Etat de la papeterie d'Etienne Montgolfier."
4. ADA, L 803, "Etat de la papeterie de Vidalon-le-Bas."
5. ADA, Q 527, document 57.
6. JO, October 17, 1789.
7. "Etat de la papeterie de Vidalon-le-Bas."
8. AN, F^{12} 1484, "Tableau des papeteries du district de Clermont."
9. "Etat de la papeterie d'Etienne Montgolfier"; and "Etat de la papeterie de Vidalon-le-Bas."
10. These figures were compiled from AN, 131 MI 53 AQ 95, 96, 97, and 98, "Journal général"; the quoted phrase is from "Etat de la papeterie de Vidalon-le-Bas."
11. Gillispie, *Montgolfier Brothers*, 131–32.
12. Gillispie, *Montgolfier Brothers*, 129–32.
13. Gillispie, *Montgolfier Brothers*, 135–36.
14. ADA, L 820, document 45, "Etat de la Compagnie de Vidalon."
15. AN, 131 MI 53 AQ 31, letter to M.-A. Montgolfier, October 23, 1798.
16. AN, 131 MI 53 AQ 31, letter to M.-A. Montgolfier, November 4, 1798.

17. ADA, L 803, document 3, decree by the National Convention of 23 nivôse, year II.
18. ADA, L 803, document 3, decree by the National Convention of 23 nivôse, year II.
19. ADA, L 803, document 35, decree of 16 fructidor, year IV.
20. Coleman, *British Paper Industry*, 192, n. 2.
21. *La Grande Encyclopédie*, 15 (1971), s.v. "Papier," 9045.
22. Quoted in Gachet, "Grèves d'ouvriers papetiers," 140.
23. Lacroix, *Historique de la papeterie*, 54.
24. André, *Machines à papier*, 107, Table 4.
25. Quoted in André, *Machines à papier*, 47, n. 25.
26. Rostaing, *Famille de Montgolfier*, 124.

CONCLUSION

1. AN, 131 MI 53 AQ 23, documents 45 and 47.

Note on Sources

This book is based primarily on the archives of the Canson and Montgolfier paper mill. Although the original records have been preserved and are available to researchers, much of the collection was filmed (with the exception of the bulky *Grands Livres*) and is accessible at the Archives nationales (Paris) under the title "Archives Canson-Montgolfier," 131 MI, 53 AQ 1 through 53 AQ 270. Administrative documents, particularly in the Archives départementales de l'Ardèche (Privas) and the Archives départementales de l'Hérault (Montpellier), Series C, proved invaluable.

A general history of premechanized French papermaking remains to be written. On the mechanized era, see the encyclopedic study by Louis André, *Machines à papier: Innovation et transformations de l'industrie papetière en France, 1798–1860* (Paris: Editions de l'Ecole des hautes études en sciences sociales, 1996). There are many local studies of French papermaking in the eighteenth century but none as sophisticated as Pierre Claude Reynard, "La Papeterie ambertoise au XVIIIe siècle: Une Prospérité fragile et stérile" (Ph.D. diss., York University, 1994). On labor relations in the same region, see Henri Gazel, *Les Anciens Ouvriers papetiers d'Auvergne* (Clermont-Ferrand: A. Dumont, 1910). There is valuable material in Robert Darnton, *The Business of Enlightenment: A Publishing History of the Encyclopédie, 1775–1800* (Cambridge: Harvard University Press, 1979), ch. 5. On the place of the Montgolfiers in the trade and much more, see Marie-Hélène Reynaud, *Les Moulins à papier d'Annonay à l'ère pré-industrielle, les Montgolfier et Vidalon* (Annonay: Editions du Vivarais, 1981). Charles Gillispie demonstrated a sure grasp of the Montgolfiers and their world in his intimate portrait *The Montgolfier Brothers and the Invention of Aviation* (Princeton: Princeton University Press, 1983). Three valuable studies for comparative purposes are D. C. Coleman, *The British Paper Industry, 1495–1860: A Study in Industrial Growth* (Oxford: Clarendon Press, 1958); Richard Hills, *Papermaking in Britain, 1488–1988: A Short History* (London: Athlone Press, 1988); and Judith McGaw, *Most Wonderful Machine: Mechanization and Social Change in Berkshire Paper Making, 1801–1885* (Princeton: Princeton University Press, 1987).

The literature on the connections between the French state and industry is voluminous and cannot be summarized here. Above all, *dirigisme*, the unyielding tutelage of the state, has been reconsidered. Among the revisionist works that have influenced this study are Jean-Louis Bourgeon, "Colbert et les corporations: L'Exemple de Paris," in *Un Nouveau Colbert*, ed. Roland Mousnier (Paris: SEDES/CDU, 1985),

242–53; Jean-Pierre Hirsch and Philippe Minard, "'Laissez-nous faire et protégez-nous beaucoup': Pour une histoire des pratiques institutionnelles dans l'industrie française (XVIIIe–XIXe siècle)," in *La France n'est-elle pas douée pour l'industrie?* ed. Louis Bergeron and Patrice Bourdelais (Paris: Belin, 1998), 135–58; Gail Bossenga, *The Politics of Privilege: Old Regime and Revolution in Lille* (Cambridge: Cambridge University Press, 1991); and Pierre Deyon and Philippe Guignet, "The Royal Manufactures and Economic and Technological Progress in France before the Industrial Revolution," *Journal of European Economic History* 9 (1980): 611–32. Additional valuable work on the French state, industry, and the economy as a whole include Eric Brian, *La Mesure de l'Etat: Administrateurs et géomètres au XVIIIe siècle* (Paris: Albin Michel, 1994); Christopher Johnson, *The Life and Death of Industrial Languedoc, 1700–1920: The Politics of Deindustrialization* (Oxford: Oxford University Press, 1995); Jean-Pierre Hirsch, *Les Deux Rêves du Commerce: Entreprise et institution dans la région lilloise, 1780–1860* (Paris: Editions de l'Ecole des hautes études en sciences sociales, 1991); Philippe Minard, *La Fortune du colbertisme: Etat et industrie dans la France des Lumières* (Paris: Fayard, 1998); and Alain Plessis, ed., *Naissance des libertés économiques* (Paris: Institut d'histoire de l'industrie, 1993).

My approach to the interplay of state, science and technology, and capital in the Old Regime was informed by Charles Gillispie, *Science and Polity in France at the End of the Old Regime* (Princeton: Princeton University Press, 1980); Roger Hahn, *The Anatomy of a Scientific Institution: The Paris Academy of Sciences, 1666–1803* (Berkeley: University of California Press, 1971); Alphonse Dupront, *Qu'est-ce que les Lumières?* (Paris: Gallimard, 1996), ch. 5; Liliane Hilaire-Pérez, "Invention and the State in Eighteenth-Century France," *Technology and Culture* 32 (1991): 911–31; and the spare sketch in Margaret Jacob, *Scientific Culture and the Making of the Industrial West* (Oxford: Oxford University Press, 1997).

In 1949, David Landes published an influential—and controversial—essay, "French Entrepreneurship and Industrial Growth in the Nineteenth Century," *Journal of Economic History* 9:45–61. Both in the realms of labor discipline and in the acquisition of up-to-date technology, the Montgolfiers were more venturesome than Landes would have us believe. See also Landes, *The Unbound Prometheus: Technological Change and Industrial Development in Western Europe from 1750 to the Present* (Cambridge: Cambridge University Press, 1969), chs. 2–3. Still worthwhile is Fernand Braudel and Ernest Labrousse, eds., *Histoire économique et sociale de la France, 1660–1789*, 2 (Paris: Presses universitaires de France, 1970). A valuable study with a contemporary emphasis is Jean-Yves Grenier, *L'Economie d'Ancien Régime: Un Monde de l'échange et de l'incertitude* (Paris: Albin Michel, 1996). François Crouzet, *Britain Ascendant: Comparative Studies in Franco-British Economic History*, trans. Martin Thom (Cambridge: Cambridge University Press, 1990), is indispensable. See also Denis Woronoff, *Histoire de l'industrie en France du XVIe siècle à nos jours* (Paris: Seuil, 1994).

On the issue of labor discipline, one must begin with the modern classics. Harry Braverman, *Labor and Monopoly Capital: The Degradation of Work in the Twentieth Century* (New York: Monthly Review Press, 1974), offers a mechanical account of the

process of deskilling. Among the most insightful studies are Michelle Perrot, "The Three Ages of Industrial Discipline in Nineteenth-Century France," in *Consciousness and Class Experience in Nineteenth-Century Europe*, ed. John Merriman (New York: Holmes and Meier, 1979), 149–68; idem, "On the Formation of the French Working Class," in *Working-Class Formation: Nineteenth-Century Patterns in Western Europe and the United States*, ed. Ira Katznelson and Aristide Zolberg (Princeton: Princeton University Press, 1986), 71–110; Alain Cottereau, "The Distinctiveness of Working-Class Cultures in France, 1848–1900," in *Working-Class Formation*, ed. Katznelson and Zolberg, 111–54; idem, "Vie quotidienne et résistance ouvrière à Paris en 1870," introduction to the reprint of Denis Poulot, *Question sociale: Le Sublime* (Paris: Maspéro, 1980); Sidney Pollard, *The Genesis of Modern Management: A Study of the Industrial Revolution in Great Britain* (Cambridge: Harvard University Press, 1965); and E. P. Thompson, "Time, Work-Discipline and Industrial Capitalism," in his *Customs in Common* (New York: New Press, 1991), 352–403.

For the history of work, see Thomas Safley and Leonard Rosenband, eds., *The Workplace before the Factory: Artisans and Proletarians, 1500–1800* (Ithaca: Cornell University Press, 1993); Daryl Hafter, ed., *European Women and Preindustrial Craft* (Bloomington: Indiana University Press, 1995); Patrick Joyce, ed., *The Historical Meanings of Work* (Cambridge: Cambridge University Press, 1987); and Steven Kaplan and Cynthia Koepp, eds., *Work in France: Representations, Meaning, Organization, and Practice* (Ithaca: Cornell University Press, 1986). Michael Sonenscher, *Work and Wages: Natural Law, Politics, and the Eighteenth-Century French Trades* (Cambridge: Cambridge University Press, 1989), is a brilliant, audacious book. Although I take issue with his assumptions about the dilution of skill in the eighteenth-century crafts, his study has set the agenda for future work on artisans and production in Old Regime France.

On the challenges endured by French workers and what they did with them, see William Reddy, *The Rise of Market Culture: The Textile Trade and French Society, 1750–1900* (Cambridge: Cambridge University Press, 1984); Joan Scott, *The Glassworkers of Carmaux: French Craftsmen and Political Action in a Nineteenth-Century City* (Cambridge: Harvard University Press, 1974); Donald Reid, *The Miners of Decazeville: A Genealogy of Deindustrialization* (Cambridge: Harvard University Press, 1985); William Sewell Jr., *Work and Revolution in France: The Language of Labor from the Old Regime to 1848* (Cambridge: Cambridge University Press, 1980); Rolande Trempé, *Les Mineurs de Carmaux, 1848–1914* (Paris: Editions Ouvrières, 1971); and Christopher Johnson, "Patterns of Proletarianization: Parisian Tailors and Lodève Woolens Workers," in *Consciousness and Class Experience*, ed. Merriman, 65–84. See also Herbert Gutman, *Power and Culture: Essays on the American Working Class*, ed. Ira Berlin (New York: Pantheon Books, 1987).

Studies of French crafts and industries in the eighteenth century are abundant. Some key titles are Serge Chassagne, *Oberkampf: Un Entrepreneur capitaliste au siècle des Lumières* (Paris: Aubier-Montaigne, 1980); Robert Darnton, "A Printing Shop Across the Border," in his *The Literary Underground of the Old Regime* (Cambridge:

Harvard University Press, 1982), 148–66; idem, *The Great Cat Massacre and Other Episodes in French Cultural History* (New York: Basic Books, 1984); Maurice Hamon and Dominique Perrin, *Au Coeur du XVIIIe siècle industriel: Condition ouvrière et tradition villageoise à Saint-Gobain* (Paris: Editions P.A.U., 1993); Michael Sonenscher, *The Hatters of Eighteenth-Century France* (Berkeley: University of California Press, 1987); Denis Woronoff, *L'Industrie sidérurgique en France pendant la Révolution et l'Empire* (Paris: Editions de l'Ecole des hautes études en sciences sociales, 1984); and the special issue of *Revue du Nord*, Gérard Gayot and Jean-Pierre Hirsch, eds., *La Révolution française et le développement du capitalisme*, 1989. On the transfer of technology, see Edward Allen, "Business Mentality and Technology Transfer in Eighteenth-Century France: The Calandre Anglaise at Nîmes, 1752–1792," *History and Technology* 8 (1990): 9–23; and John Harris, *Essays in Industry and Technology in the Eighteenth Century: England and France* (Brookfield, Vt.: Variorum, 1992). For a valuable comparison, see Richard Goldthwaite, *The Building of Renaissance Florence: An Economic and Social History* (Baltimore: Johns Hopkins University Press, 1980).

The debate over consumption and its role in stimulating industrial change has grown lively in recent years. Two fine entry points are John Brewer and Roy Porter, eds., *Consumption and the World of Goods* (New York: Routledge, 1993); and Colin Jones, "Bourgeois Revolution Revivified: 1789 and Social Change," in *Rewriting the French Revolution*, ed. Colin Lucas (Oxford: Clarendon Press, 1991), 69–118. See also Leora Auslander, *Taste and Power: Furnishing Modern France* (Berkeley: University of California Press, 1996).

Finally, the eighteenth century was the age of the technical manual par excellence, and papermaking was no exception. Among the many available sources, see Nicolas Desmarest, "Papier (Art de fabriquer le)," *Encyclopédie méthodique: Arts et métiers mécaniques*, 5 (Paris, 1788).

Index

Library of Congress Cataloging-in-Publication Data

Rosenband, Leonard N.

Papermaking in eighteenth-century France : management, labor, and revolution at the Montgolfier Mill, 1761–1805 / Leonard N. Rosenband.

p. cm.

Includes bibliographical references and index.

ISBN 0-8018-6392-9 (alk. paper)

1. Paper industry—France—Annonay—History. 2. Industrial relations—France—Annonay—History. 3. Montgolfier family.

HD9832.8.A56 R67 2000

338.7′676′0944589—dc21 99-086391

D1116983